Corrosion Inhibitors

Edited by Ambrish Singh

Published in London, United Kingdom

IntechOpen

Supporting open minds since 2005

Corrosion Inhibitors
http://dx.doi.org/10.5772/intechopen.76742
Edited by Ambrish Singh

Contributors
Ambrish Singh, Kashif Ansari, Yuanhua Lin, Mumtaz Quraishi, Geethamani P, Funsho Olaitan Kolawole, Shola Kolawole, Oluwamayowa Olugbemi, Suleiman Hassan, Pavel Topala, Alexandr Ojegov, Besliu Vitalie, Mohammad Mazumder, Lipiar Khan, Kannan Perumal, Olusola Solomon Amodu, Moradeyo Odunlami, Joseph Taiwo Akintola, Seide Akoro, Seteno Karabo Obed Ntwampe, Abdenacer Berradja

First published in London, United Kingdom, 2019 by IntechOpen
IntechOpen is the global imprint of INTECHOPEN LIMITED, registered in England and Wales, registration number: 11086078, The Shard, 25th floor, 32 London Bridge Street
London, SE19SG – United Kingdom
Printed in Croatia

British Library Cataloguing-in-Publication Data
A catalogue record for this book is available from the British Library

Additional hard and PDF copies can be obtained from orders@intechopen.com

Corrosion Inhibitors
Edited by Ambrish Singh
p. cm.
Print ISBN 978-1-78984-714-7
Online ISBN 978-1-78984-715-4
eBook (PDF) ISBN 978-1-78985-569-2

For EU product safety concerns:
IN TECH d.o.o., Prolaz Marije Krucifikse Kozulić 3, 51000 Rijeka, Croatia, info@intechopen.com or visit our website at intechopen.com.

Meet the editor

Dr. Ambrish Singh is a professor in the School of Materials Science and Engineering, Southwest Petroleum University, China. He leads the Corrosion Research Group with master's students, PhD students, and postdoctoral researchers. His research interests are mainly focused on corrosion, electrochemistry, green chemistry, quantum chemistry, smart coatings, polymers, nanomaterials, composites, and petroleum engineering. He was invited to the Indian Science Congress 2019 as the only foreign speaker on chemical sciences. Dr. Singh received the prestigious Sichuan 1000 Talent Award from the Sichuan government, China, for his outstanding research contributions as a faculty, and the President's Award for exceptional postdoctoral research work. He also received the Young Scientist Award from UPCST, Lucknow, India. He has published more than 100 SCI peer-reviewed research papers in high-impact journals and acts as a reviewer for more than 40 high-impact journals and editor for a small number of journals. He is included as board member in several journals due to his contributions to the field of corrosion. Dr. Singh has been invited to present his work in several national and international conferences, seminars, and workshops. He has drafted five patents and filed two patents based on his innovative findings in China. He has finished several state and provincial projects in China and India. He is the member of the National Association for Corrosion Engineers, Houston, USA, the Society of Petroleum Engineers, and the American Chemical Society. He collaborates with other corrosion groups in Germany, South Korea, the United Kingdom, Portugal, Italy, and South Africa.

Contents

Preface

Progress in the field of corrosion mitigation has been commendable in recent years. As the use of metals and alloys increased throughout the world, the problem of corrosion increased simultaneously. The available methods do not provide ample solutions to the existing problems. So, the area is wide open to explore and develop further options and provide suitable mitigation to the corrosion of metals and alloys. Among the active methods, use of inhibitors is common and is used globally due to ease of usage, availability, and low cost.

The strict environmental guidelines and measures do not allow the usage of organic inhibitors in high concentrations. Therefore, the development of inhibitors is mostly focused on green principles and eco-friendly nature. The inhibitors prepared following the green principles are environmentally benign and serve as a potential solution to mitigate corrosion in petroleum industries and refineries. Different inhibitors have been reported with several formulations for industries to delay the corrosion process. To explain the mechanism of inhibitor action, several examples have been included with appropriate equations for different metals in different aggressive media. Detailed emphasis has been given to determine the corrosion rate using various available techniques. Several chapters have been included in this book to elaborate the mechanisms of inhibitor action in different corrosive solutions.

Apart from traditional studies, quantum studies have been used extensively to support the experimental findings. The introduction of density functional theory has strongly influenced the explanations of mechanisms and surface phenomena. Basic concepts such as chemical potential, chemical hardness, electronegativity, electrophilicity, and nucleophilicity play important roles in the prediction of reaction mechanisms and the analysis of chemical reactions. One chapter written by Dr. Ambrish Singh has been included that mainly focuses on the use of quantum studies in corrosion inhibition.

Tribocorrosion is one of the hot topics in corrosion research that involves materials and mechanical aspects altogether. Tribocorrosion damage (i.e., material loss) can be designated in a broad sense as a failure mechanism due to the mutual interaction of corrosion, friction, and wear processes and their synergy effects. It generates changes in surface and/or volume compositions, often modifies the environment, and ultimately can lead to system failure. The chapters by Dr. Abdenacer Berradja include overall insights into tribocorrosion in this book.

Finally, I would like to thank all the committee members of IntechOpen along with all authors and reviewers for their invaluable contributions.

Dr. Ambrish Singh
School of Materials Science and Engineering,
Southwest Petroleum University,
Chengdu, Sichuan, China

State Key Laboratory of Oil and Gas Reservoir Geology and Exploitation,
Southwest Petroleum University,
Chengdu, Sichuan, China

Section 1

Inhibitors

Chapter 1

Corrosion Inhibitors

Geethamani Palanisamy

Abstract

Corrosion is a natural process driven by energy consideration. Inhibition is a preventive measure against corrosive attack on metallic materials. Corrosion inhibitors have been frequently studied, since they offer simple solution for protection of metals against corrosion in aqueous environment. Mineral acids like hydrochloric and sulfuric acids are most widely used in pickling baths to remove the metal oxides formed on the surface. The multidisciplinary aspect of corrosion problems combined with the distributed responsibilities associated with such problems only increase the complexity of the subject. Inhibitors are used in industrial and commercial processes to minimize both the metal loss and acid consumption.

Keywords: corrosion inhibitors, acidic inhibitors, volatile inhibitors, vapor phase inhibitor

1. Introduction

Corrosion is the primary means by which metals deteriorate. Corrosion introduces itself into many parts of our lives [1, 2]. The great majority of us have personal feeling for the importance of corrosion. Far too many have cringed at the emergence of rust holes in the body panels of relatively new automobiles [3]. The outdoor rusting of steel, household and garden appliances is a common fact of life. All have seen the strains on cooking utensils from hot foods or experienced the metallic taste in acid foods stored too long on open cans. That these effects are caused by corrosion is well known [4]. The glaring example related to corrosion is the appearance of cracks in certain portions of Taj Mahal, was due to steel dowels embedded inside had extensively corroded and rusted leading to fractures in the stoned [5]. However, corrosion is just as common in other material classes such as ceramics, plastics and rubber. Since, practically all environments are corrosive to some degree and are major contributing causes of material failure and also are a large economic cost to the society [6].

2. Historical background

Corrosion can be viewed as a universal phenomenon, omnipresent and omnipotent. It is there everywhere, air, water, soil and in every environment, we encounter [7]. Known to people as rust, corrosion is an undesirable phenomenon which destroys the luster and beauty of the materials and lessens their life. Indian government spending around 3.5% Lakscrores of the nation's GDP per annum for losses of corrosion [8]. Recent studies estimate that, not only in India, other countries also rise their funds for demand of corrosion inhibitors [9].

Corrosion costs manifest in the form of premature deterioration or failure necessitating maintenance, repairs and replacement of damaged parts.. Corrosion has a vast environmental and economic impact on all the surfaces of national infrastructure like highways, bridges, buildings, chemical processing units, waste water treatment and virtually on all metallic objects in our day to day life use [10]. Other than material loss, corrosion interferes not only with environment, also affects human safety and industrial operations severely. Awareness to corrosion and adaptation of timely and appropriate control measures hold the key in the abatement of corrosion failures [11].

3. Corrosion definition

The spontaneous oxidation of metal is termed as corrosion [12], that is, Corrosion is the deterioration or destruction of metals and alloys in the presence of an environment by chemical or electrochemical means. The medium in which the metal undergoes corrosion is termed as corrosive or aggressive medium. Corrosion products formed are chemical compounds containing the metal in the oxidized form with the exception of gold and platinum, all other metals corrode and transform themselves into substances similar to the mineral ores from which they are extracted [13].

4. Adverse economic and social effects of corrosion

The corrosion affects severely on the safe, reliable and efficient operation of equipment and structures than the simple loss of a mass of metal [14]. Failures of all kind of machineries and the need for expensive replacements may occur even though the amount of metal destroyed is quite small. Some of the major harmful effects of corrosion can be listed below:

Nuclear plant shutdown due to failure, for example, nuclear reactor during decontamination process.

- Replacement of corroded equipment resulting in heavy expenditure.
- High cost preventive maintenance such as painting.
- Loss of efficiency.
- Loss of product from a corroded container.
- Safety requirement measures from a fire hazard or explosion or release of toxic product.
- Health problems, for example, drinking water contamination with lead is likely due to corrosion.

5. Chemistry of corrosion

In general, metals are having unique properties like opaque, lustrous, conductivity, malleable and ductile in nature and are readily forms metallic bonds with other metals and ionic bonds with non-metals [15, 16]. The metals that have overlapping conduction bands and valence bands in their electronic structure.

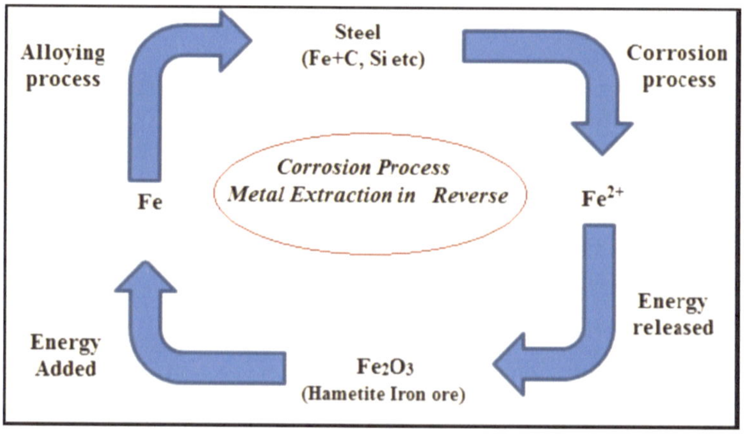

Figure 1.
Corrosion cycle process.

Metals are obtained from their ore by the expenditure of large amounts of energy. Metals store heat as potential energy during the smelting and refining process and release this energy during the corrosion process after reacting with the environment. These metals can therefore be regarded as being in a metastable state and will tend to lose their energy by reverting to compounds more or less similar to their original states, for example the starting material for iron and steel making and the corrosion product rust has the same chemical composition (Fe_2O_3).

The energy stored during melting and released during corrosion supplies the driving potential for the corrosion process to take place. Since most metallic compounds, and especially corrosion products, have little mechanical strength, a severely corroded piece of metal is quite useless for its original purpose [17]. Metals such as Mg, Al, Zn, and Fe which require larger amount of energy for refining are more susceptible to corrosion than metals which require lesser amount for refining such as gold, silver, platinum. A corrosion cycle is shown below (**Figure 1**).

6. Classification

Corrosion has been classified into different methods. They are

- Low temperature corrosion and high temperature corrosion (or)
- Electrochemical corrosion and chemical corrosion (or)
- Wet and Dry corrosion.

Wet corrosion occurs when the metal is in contact with an electrolytic conducting liquid or when two dissimilar metals or alloys are either immersed or dipped partially in the electrolytic conducting solutions. This is always associated with low temperature conditions. The corrosion process involves two reactions.

At anode:

M (metal) $\rightarrow Mn^{+} + ne^{-}$ (oxidation)
Mn + (metal ion) $\rightarrow$ Dissolves in solution
Mn + (metal ion) $\rightarrow$ Forms compound such as oxide

At Cathode:

$$2H^+ + 2e^- \rightarrow H_2\uparrow \text{ (in acid solution)}$$
$$2H_2O + 2e^- \rightarrow H_2\uparrow + 2OH^- \text{ (in alkaline solution)}$$
$$O_2 + 2H_2O + 4e^- \rightarrow 4OH^- \text{ (in neutral solution)}$$

Dry corrosion takes place mainly through the direct chemical action of atmospheric gases and vapors present in the environment 30. This is most often associated with high temperature.

7. Forms of corrosion

Corrosion can manifest itself in many forms such as uniform corrosion or general corrosion, galvanic corrosion, crevice corrosion, pitting corrosion, intergranular corrosion, selective leaching, erosion corrosion, stress corrosion, corrosion fatigue and fretting corrosion [18]. In order to improve the understanding between corrosion and design engineers it is classified into two broad categories. They are expressed in flow chart (**Figure 2**).

7.1 General corrosion

This general corrosion also called as a uniform attack is the most common form of corrosion. It is normally characterized by a chemical or electrochemical reaction which proceeds uniformly over the entire exposed surface or over a large area [19]. The metal becomes thinner and eventually fails.

7.2 Galvanic corrosion

It occurs when a potential difference exists between two dissimilar metals immersed in a corrosive solution. This potential difference produces a flow of electrons between the metals. Several investigations have shown that, galvanic corrosion is directly proportional to the area of the cathodic to the anodic metal [20]. A schematic diagram for galvanic corrosion is shown below(**Figure 3**).

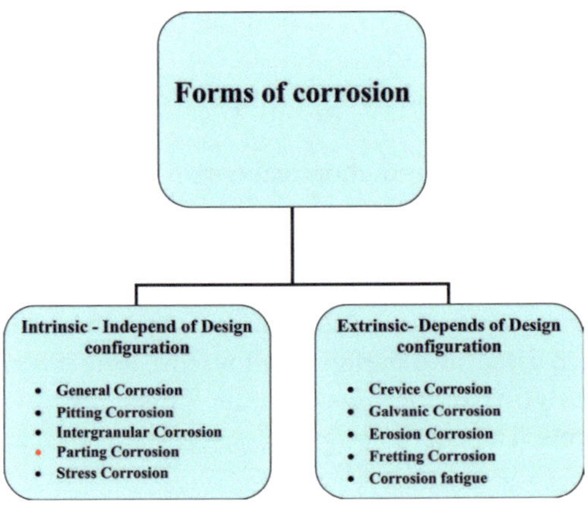

Figure 2.
Forms of corrosion.

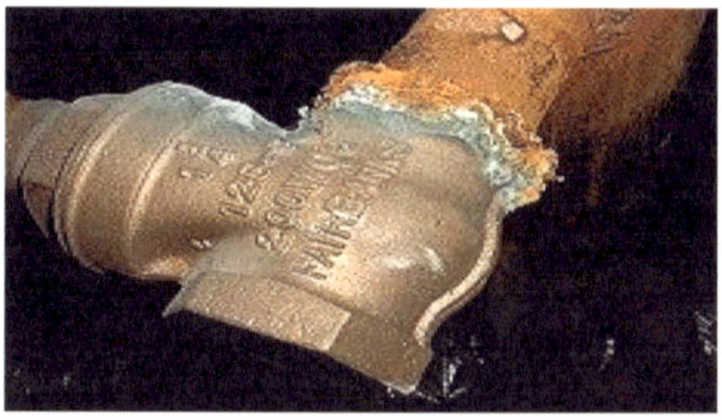

Figure 3.
Schematic representation of galvanic corrosion.

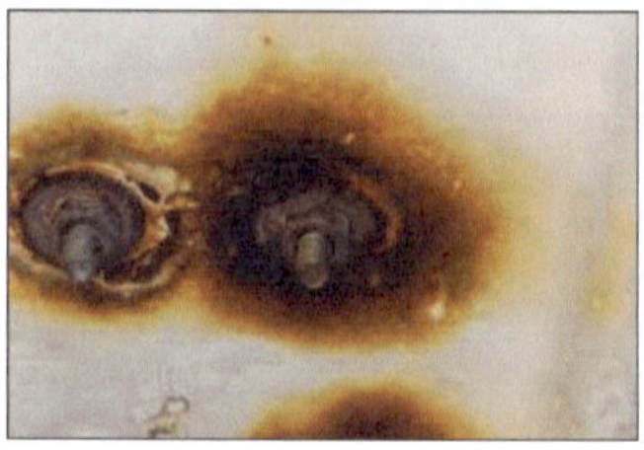

Figure 4.
Schematic representation of crevice corrosion.

7.3 Crevice corrosion or deposit corrosion

This kind of corrosion is attacked generally within crevices associated with small volumes of stagnant solution trapped in holes, surfaces, joints and crevices under bolt and rivet heads (**Figure 4**) [21]. It is also known as deposit or gasket corrosion.

7.4 Pitting corrosion or localized corrosion

Pitting corrosion is a localized attack resulting in the formation of holes in the metals. These holes are relatively small and they are looks like a rough surface (**Figure 5**), they were sometimes isolated or so close together. Pitting is one of the most destructive and insidious forms of corrosion [22].

7.5 Intergranular corrosion

Most of the metals and alloys are susceptible to intergranular corrosion, when exposed to specific corrosion environment which is shown in the **Figure 6**. Grain boundaries are usually more reactive than grain matrix. Hence localized attack occurs at and adjacent to grain boundaries with relatively little corrosion of the matrix. This type of attack is usually rapid and penetrates deep into the metal resulting in loss of strength and causes catastrophic failures. It is caused by,

- Impurities at the grain boundaries
- Enrichment of one of the elements in the alloy
- Depletion of one of the elements in the boundary area.

7.6 Selective leaching

Selective leaching is the removal of an element from an alloy by corrosion. Selective removal of zinc from brass is a prime example of this form of attack. A similar attack has been observed with other alloys in which iron, aluminum, cobalt and chromium are removed. This type corrosion is undesirable as it yields a porous metal with poor mechanical properties (**Figure** 7).

7.7 Erosion corrosion

It is the increase of attack of a metal because of relative movement between a corrosive medium and the metal surface. This type of erosion corrosion is usually associated with systems where high velocities of corrosive fluids or gases are encountered.

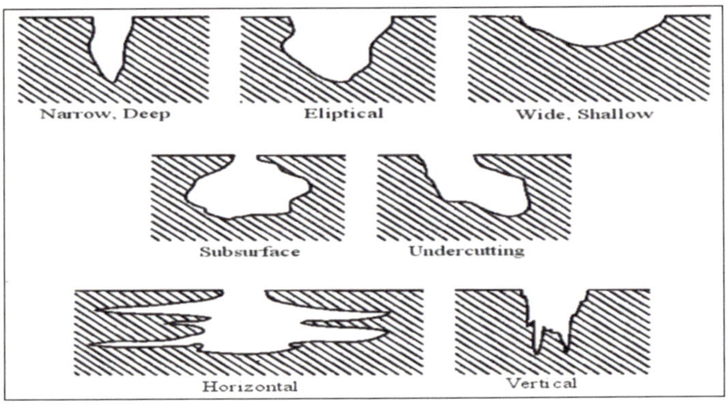

Figure 5.
Schematic representation of pitting corrosion.

Figure 6.
Schematic representation of intergranular corrosion.

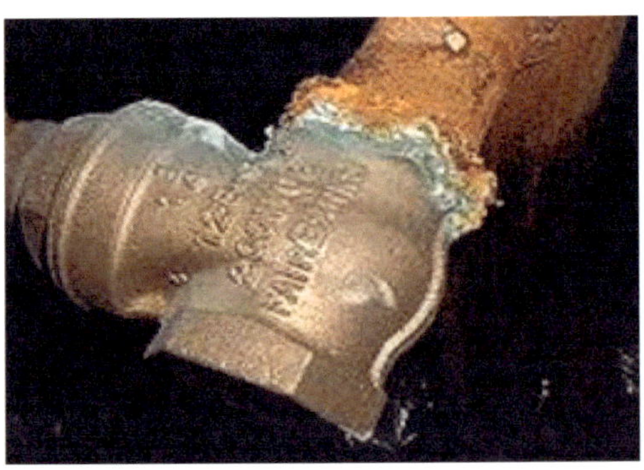

Figure 7.
Schematic representation of selective leaching corrosion.

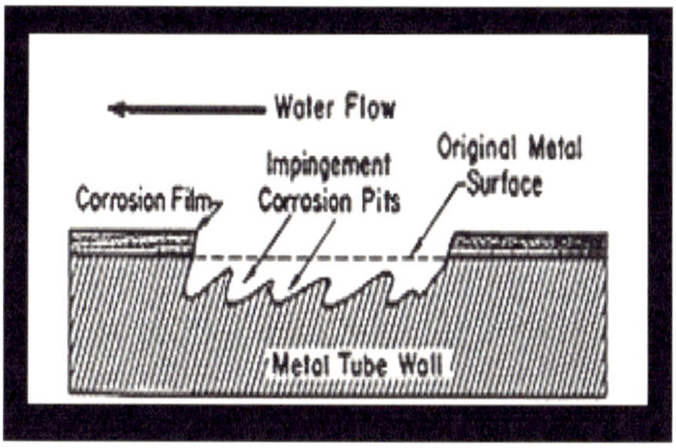

Figure 8.
Schematic representation of erosion corrosion.

This corrosion can be observed in piping system such as bends, elbows, pumps and condensers, etc. A Schematic representation of erosion corrosion is shown in **Figure 8**. Factors affecting erosion corrosion are nature of surface film, corrosion environment and presence of air bubbles with its size, chemical composition, suspended solids, corrosion resistance and metallurgical properties of metals and alloys.

7.8 Stress corrosion

The cracking of metal or alloy by the combined action of a tensile stress and a corrodent (**Figure 9**) is known as stress corrosion cracking. The susceptibility to stress corrosion cracking is due to certain metallurgical factors such as,

- Chemical composition
- Preferential orientation of grains
- Composition and distribution of precipitates
- Dislocation structure and environmental factors and structure of metal.

7.9 Corrosion fatigue

It is defined as, "the reduction of the fatigue strength due to the presence of corrosive environment". Corrosion fatigue occurs due to the combined action of tensile and compressive stress alternatively. Fatigue occurs at lower stress in corrosive environment.

Figure 9.
Schematic representation of stress corrosion.

7.10 Fretting corrosion

Fretting is a wear phenomenon enhanced by corrosion. It involves wear of a metal or alloy when in contact with another solid material in dry or humid air. Fretting is the result of abrasive wear surface oxide films, which form a contacting surfaces under load in atmospheric air which is shown in **Figure 10**. The factors which affect fretting corrosion are:

- Magnitude of relative motion
- Temperature
- Environment
- Metallurgical factors

Due to slight motion such as vibration, surface oxide and underlying metal, gets spoiled. The metal particles as a result of wear get oxidized to hard oxides which act as an additional abrasive medium. Further, the motion grinds the oxides particles thus causing wear.

Figure 10.
Schematic representation of fretting corrosion.

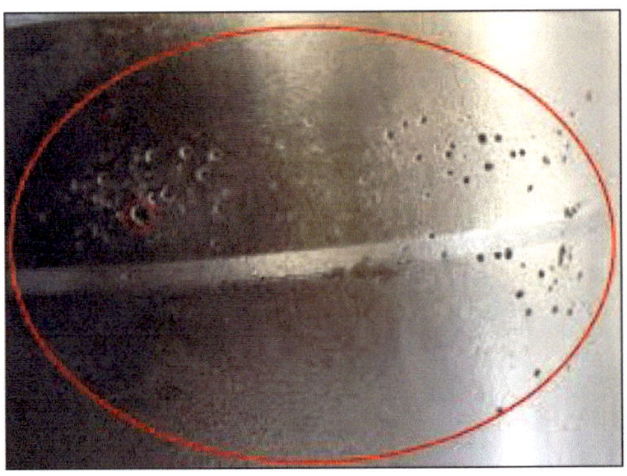

Figure 11.
Schematic representation of cavitation corrosion.

7.11 Cavitations corrosion

It is a special type of erosion corrosion which is caused due to the formation of vapor bubbles in a corrosive environment near a metal surface and when the bubbles collapse, attack arises for example, hydraulic turbulence, ship propellers, etc. It is similar to pitting corrosion but the surface is rough and has many close spaced pits (**Figure 11**).

8. Factors influencing corrosion

The extent and rate of corrosion depend on nature of the metals and the environments.

8.1 Nature of the metal

1. Position of metals in EMF series
2. Overvoltage
3. Relative area of anodic and cathodic parts of the metal
4. Purity of the metal
5. Physical nature of the metal
6. Nature of the surface film
7. Solubility of products
8. Volatile corrosion products.

8.2 Nature of the environment

- Temperature range
- Humidity of air
- Impurities in water
- Presence of suspended particles in atmosphere
- Influence of pH
- Nature of dissolved gases, dissolved salts, pollutants, etc.
- Conductance of the corroding medium
- Formation of oxygen concentration cell
- Flow velocity of process steam
- Polarization of electrodes.

9. Factors controlling corrosion rate

Certain factors tend to accelerate the action of a corrosion cell which includes the establishment of well-defined locations on the metal surface for the anodic and cathodic reactions. Metals having a more positive (noble) potential in the galvanic series will tend to extract electrons from a metal which is in a more negative (base) position in the series and hence accelerate its corrosion when in contact with it [23]. Aggressive ions such as chloride tend to prevent the formation of protective oxide films on the metal surface and thus increase corrosion.

9.1 Corrosion rate

The rate of corrosion is expressed based on the loss per unit time. The rate at which the attack takes place is of prime importance and is usually expressed in one of the two ways:

- Weight loss per unit area per unit time, usually mdd (milligrams per square decimeter per day).
- Decrease in thickness per unit time, that is, rate of penetration or the thickness of metal lost. This may be expressed in American units, mpy (mils per year) or in metric units or mmpy (millimeters per year).

10. Corrosion control methods

Corrosion is destructive and silent operating processes. It poses problems to big as well as small industries. Since corrosion is inevitable to eliminate but can be minimized by adopting certain anticorrosion method rather than preventing it [24]. The practical methods available for the protection of metal against corrosion are diverse. They may be broadly based on,

- Modification of metal
- Modification of design
- Modification of corrosive environment
- Modification of metal environment potential
- Use of inhibitors
- Modification of surface

These methods can be used individually or in combination.. One of the best known methods of corrosion protection is using corrosion inhibitors instead of using the various methods to avoid or prevent destruction or degradation of metal surface. Because using inhibitors is following stand up due to low cost and practice method [25, 26].

11. Inhibitors

An inhibitor is a chemical substance or combination of substances which when added in very low concentrations in a corrosive environment effectively prevents or reduces corrosion without significant reaction with the components of the environment. Concentrations of corrosion inhibitors can vary from 1 to 15,000 ppm (0.0001 to 1.5 wt %). Inhibitors play a vital role in closed environmental systems that have good circulation so that an adequate and controlled concentration of inhibitor is ensured [27]. Such conditions can be met, for instance in cooling water recirculation systems, oil production, oil refining, and acid pickling of steel components. One of the more recognizable applications for inhibitors is in antifreeze for automobile radiators. Inhibitors may be organic or inorganic compounds and they are usually dissolved in aqueous environments [28].

11.1 Definition of inhibitor

An inhibitor is a substance which when added to an environment in small concentration minimizes the loss of metal, reduces the extent of hydrogen embrittlement, protects the metal against pitting, reduces over pickling and acid fumes resulting from excessive reaction between the acid and basic metals and reduces acid consumption. They reduce corrosion by either acting as a barrier by forming an adsorbed layer or retarding the cathodic, the anodic or both processes [10]. A schematic representation of inhibitor process is shown in **Figure 12**.

Any corrosion retardation process or the reduction in the oxidation rate of the metal by addition of a chemical compound to the system is caused by corrosion inhibitors. Inhibitors are often easy to apply and offer the advantage of in-situ application without causing any significant disruption to the process. The use of corrosion inhibitors is one of the best methods of combating corrosion [11].

In order that they can be used effectively, three factors must be considered, namely:

- Identification of the corrosion problems.

Generally, the components of a corrosion cell (anode, cathode, electrolyte and electronic conductor) may be affected by corrosion inhibitors in order to reduce corrosion. The inhibitor may cause:

- Anodic inhibition (increasing the polarization of the anode)
- Cathodic inhibition (increasing the polarization of the cathode)
- Resistance inhibition (increasing the electrical resistance of the circuit while forming a thin or thick deposit on the surface of the metal)
- Diffusion restriction (restricting the diffusion of depolarizers, e.g., DO).

However there are several factors to be considered when choosing an inhibitor.

- Cost of the inhibitor.
- Toxicity of the inhibitor can cause ill effects on human beings and other living species.
- Availability of the inhibitor determines the selection of it.
- Inhibitor should be environment friendly.

In order to avoid or reduce the corrosion of metallic materials, inhibitors used in cooling system must satisfy the following criteria

- It must give good corrosion protection at a very low concentration of inhibitor.
- It must protect all exposed materials from the attack of corrosion.
- It must remain efficient in extreme operating conditions (higher temperature and velocity).
- In case of an under or over dosage of inhibitor, corrosion rate should not increase drastically.
- The inhibitor or reaction products of the inhibitor should not form any deposits on the metal surface particularly at locations where heat transfer takes place.
- It should suppress both uniform and localized corrosion.
- It should have long range effectiveness.
- It should not cause toxicity and pollution problems.

11.2 Definition of inhibition

One of the extensively studied topics in the field of corrosion is inhibition. Inhibition is a process of preventive measure against corrosive attack on metallic materials. Chemical compounds may be used which, when added in small concentrations to an aggressive environment, are able to decrease corrosion of the exposed metal [12].

11.3 Corrosion inhibitors

Corrosion inhibitors can be any forms (solids, liquids and gases). Based on the solubility or dispersibility in fluids corrosion inhibitors are selected which are to be inhibited. Corrosion inhibitors have been found to be effective and flexible means of corrosion mitigation. The use of chemical inhibitors to decrease the rate of corrosion processes is quite variable. Corrosion inhibitors are used in oil and gas exploration and production, petroleum refineries, chemical manufacturing, heavy manufacturing, water treatment and product additive

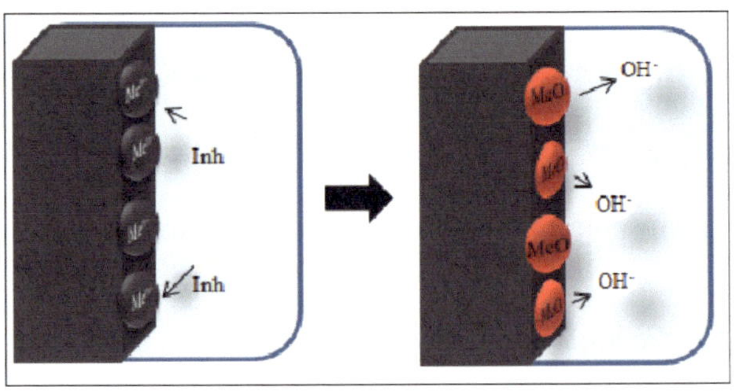

Figure 12.
Schematic representation of inhibition process.

industries [29]. In the oil extraction, processing and chemical industries, corrosion inhibitors have always been considered to be the first line of defense. A great number of scientific studies have been devoted to the subject of corrosion inhibitors [30, 31].

11.4 Role of inhibitors

Organic and inorganic compounds constitute a large class of corrosion inhibitors, which as a general rule; affect the entire surface of a corroding metal when present in sufficient concentration. Most of the organic/inorganic compounds containing elements of groups V B, VI B or functional groups of the type amine, carbonyl and alcoholic groups are more effective corrosion inhibitors. The inhibitor is adsorbed on the entire metal surface of the corroding metal and so prevents attack from the corrosion. Organic inhibitors are adsorbed on the metals surface [32]. The inhibitors may be considered as two fundamental types, they are, those which form a protective barrier film on anodes or cathodes by reaction between the metal and the environment. This type of inhibitors function in neutral or in some cases, alkaline solution in which the main cathodic reaction is an oxygen reduction reaction in which the corroding metal surface is covered by a film oxide or hydroxide.

Another type is initially adsorbed directly onto the metal surface by interaction between surface charges and ionic and/or molecular dipole charges. This division of inhibitor types results principally from the pH of the solution where they operate. Inhibitors must be present in a minimum concentration for them to be fully effective. This is very common with anodic inhibitors.

The efficiency of organic inhibitors can be improved in the presence of certain halogen ions. Halogen ions are also known to inhibit corrosion to some extent in acid solutions. The efficiency of the corrosion inhibition is in the order; I- > Br- > Cl-. Fluoride does not show inhibition characteristics. Synergism of halogen ions can be attributed to the fact that the metal adsorbs halogen ions whose charge shifts the surface in a negative direction, thereby increasing adsorption of the cationic organic inhibitor. Being able to discover possible compounds that can be used as corrosion inhibitors requires a lot of hard work, innovation and laboratory analysis/synthesis [33].

Certain halogen ions present in the organic inhibitors are known to inhibit corrosion to some extent in acid solutions. The efficiency of the corrosion inhibition is following in the order of $I^- > Br^- > Cl^-$. Fluoride does not shows any inhibition characteristics. Synergism of halogen ions can be attributed to the fact that the metal adsorbs halogen ions whose charge shifts the surface in a negative direction, thereby increasing adsorption of the cationic organic inhibitor.

11.5 Classification of corrosion inhibitors

Corrosion inhibitors are briefly classified (**Figure 13**) as follows,

11.5.1 Based on electrode process

11.5.1.1 Anodic inhibitors

An anodic inhibitor increases the anodic polarization and hence moves the corrosion potential to the cathodic direction and hence also called as passivating inhibitors. Anodic inhibitors such as chromates, phosphates, tungstates and other ions of transition elements with high oxygen content are those that stifle the

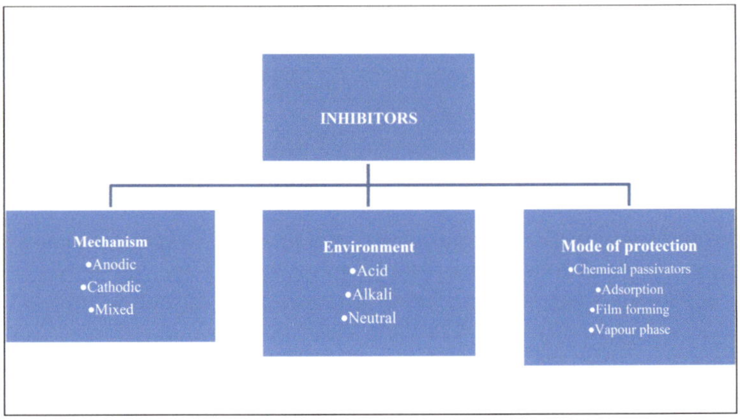

Figure 13.
Classification of corrosion inhibitors.

corrosion reaction occurring at the anode by forming a sparingly soluble compound with a newly produced metal ion. They are adsorbed on the metal surface forming a protective film or barrier, thereby reducing the corrosion rate. Anodic inhibitors build a thin protective film along the anode and increasing their potential and thus slow down the corrosion reaction [34].

Although, this type of control is affected, yet it may be dangerous since severe local attack can occur, if certain areas are left unprotected by depletion of the inhibitors. A number of inorganic inhibitors such as orthophosphates, silicates, etc. fall under anodic type. Even though anodic inhibitors are widely used, a few of them have some undesirable property. If such inhibitors are used in very low concentrations, they cause stimulation of corrosion such as pitting and for this reason anodic inhibitors are denoted as dangerous.

There are two types of passivating inhibitors.

- The oxidizing anions such as chromates, nitrites and nitrates that can passivate steel in the absence of oxygen.

- The non-oxidizing ions such as phosphates, tungstates and molybdates that require the presence of oxygen to passivate steel.

In general, passivation inhibitors can actually cause pitting and accelerate corrosion when concentrations fall below minimum limits. For this reason, it is essential to monitor the inhibitor concentration.

11.5.2 Cathodic inhibitors

Cathodic inhibitors reduce corrosion by slowing the reduction reaction rate of the electrochemical corrosion cell. This is done by blocking the cathodic sites by precipitation. Cathodic inhibitors are effective when they slow down the cathodic reaction. Elements As, Bi and Sb are referred to as cathodic poisons which reduce the hydrogen reduction reaction rate and lower the overall corrosion rate. Removal of oxygen from the corrosive environment will significantly decrease the corrosion rate. This can be done through

- The use of oxygen scavengers such as sodium sulfite and hydrazine which react with the oxygen and remove it from the solution

- Vacuum de-aeration or

- Boiling to lower the dissolved oxygen concentrations.

Cathodic inhibitors shift the corrosion potential to the anodic direction. [35, 36] Here the cations migrate towards the cathode surfaces where they are precipitated chemically or electrochemically and thus block these surfaces. The inhibiting action of cathodic inhibitors takes place by three mechanisms,

- **Cathodic poisons**: The cathodic reduction process is suppressed by impeding the hydrogen recombination and mode of protection discharge but increase the tendency of the metal to be susceptible to hydrogen induced cracking.

- **Cathodic precipitates**: Compounds such as calcium, magnesium will precipitate as oxides to form a protective layer which acts as a barrier on the metal surface.

- **Oxygen scavenger**: These compounds react with oxygen present in the system to form a product and reduce corrosion. For example, As^{3+} and Sb^{3+} on dissolution of Fe in acids.

11.5.2.1 Mixed inhibitors

These inhibitors retard both the anodic and cathodic processes involved in the corrosion process and are therefore called mixed inhibitors [37]. They are typically film forming compounds that cause the formation of precipitates on the surface blocking both anodic and cathodic sites indirectly. Anodic inhibitors are, for the most part, dangerous inhibitors, especially if their concentrations are too less. But cathodic inhibitors are generally safe. Mixed inhibitors are less dangerous than pure anodic inhibitors, and in number of cases they may not increase the corrosion intensity. The most common inhibitors of this category are the silicates and the phosphates. Such inhibitors will have the more advantage that they control both the cathodic and anodic corrosion reactions.

11.5.3 Based on environment

11.5.3.1 Acidic environment inhibitors

11.5.3.2 Inorganic inhibitors

The compounds such as As_2O_3, Sb_2O_3 have been reported as inhibitors in acid media. In this case, the protection is due to the reduction of electro positive ions and deposition on the metal surface and lowering of the over voltage of main cathodic depolarization reaction [38]. Recently it is shown that the addition of heavy metal ions such as Pb^{2+}, Ti^{+}, Mn^{2+} and Cd^{2+} is found to inhibit corrosion of iron in acids.

11.5.3.2.1 Organic inhibitors

Organic compound containing oxygen, nitrogen, sulfur with multiple bonds have been reported as good corrosion inhibitors. Many organic inhibitors such as amines, aldehydes, alkaloids, nitro and nitroso compounds have been studied and tried as corrosion inhibitors [39]. Organic inhibitors can be anodic, cathodic and mixed type based on its reaction at the metal surface and potential. These are effective depending upon its size, carbon chain length, aromaticity, conjugation and nature of bonding atoms [40].

11.5.3.3 Alkaline inhibitors

Metals, which form amphoteric oxides, are prone to corrosion in alkaline solutions. Many organic compounds are often used as inhibitors for metals in basic solutions [41]. Compounds such as thiourea, substituted phenols, naphthol, β-diketone, etc., have been used as effective inhibitors in basic solutions due to the formation of metal complexes.

11.5.3.4 Neutral inhibitors

Inhibitors which are effective in acidic solutions do not function effectively in neutral solutions, since the mechanism is different in the two solutions [42–44]. In neutral solutions, the interaction of inhibitors with oxide covered metal surface and prevention of oxygen reduction reaction at the cathodic sites takes place. Such inhibitors protect the surface layers from aggressiveness. Some surface active chelating inhibitors have been found to be efficient inhibitors in near-neutral solutions [45].

11.5.4 Based on mode of protection

11.5.4.1 Chemical passivators

Substances which usually have a sufficiently high equilibrium potential (redox or electrode potential) and sufficiently low over potential decrease corrosion rate on attainment of passivity and are called chemical passivators [46].

Example, nitrites are used as inhibitors for antifreeze cooling waters. Chromates are mostly used as inhibitors for recirculating cooling waters. Zinc molybdate is used as an inhibiting pigment for paints.

11.5.4.2 Adsorption inhibitors

These represent the most widely used class of inhibitors. In general, they are organic compounds which get adsorbed on the metal surface and provide a blanketing effect over the entire surface, that is, both in cathodic and anodic cases. Generally they effect both cathodic and anodic reactions equally, but in many cases the effect may not be equal. These are commonly used in the acid pickling of hot rolled products in order to remove the black mill scale and are thus known as pickling inhibitors [47–49].

Examples: Compounds containing lone pairs of electrons such as nitrogen atoms in amines, quinolines, sulfur atoms in thio compounds and oxygen atoms in aldehydes.

11.5.4.3 Film forming inhibitors

In contrast to the adsorption inhibitors which form the straight forward adsorbed film of the inhibiting species, many substances called film forming inhibitors, appear to stop corrosion by forming a blocking or a barrier film of a material other than the actual inhibiting species itself. Such materials tend to be specific either to the cathode or to the anode. Zinc and calcium salts are the most common examples of cathodic film forming inhibitors. Benzoate is the common example of anodic film forming inhibitors, which inhibit corrosion during voyages [50].

11.5.4.4 Vapor phase inhibitors

Atmospheric corrosion of metals in closed spaces as in parcels during storage and shipment can be prevented by the use of certain substances called vapor phase inhibitors also called as volatile inhibitors [51–56]. These are substances of low but significant vapor pressure. The vapor comes in contact with the surface of the metal and the adsorption of the inhibitor takes place. The moisture then hydrolyses it and releases protective ions which have corrosion inhibiting properties. **Figure 14** shows the schematic representation vapor phase inhibitors.

Examples,

Dicyclohexylamine chromate and benzotriazole for protecting copper.

Phenyl thiourea and cyclohexylamine chromate for brass.

Dicyclohexylamine nitrite for ferrous and nonferrous part.

11.5.4.5 Volatile inhibitors or vapor phase inhibitors

Volatile corrosion inhibitors (VCIs) are compounds which transferred in a closed environment to the site of corrosion by volatilization from a source **(Figure 15).** If the corrosion product is volatile, it volatilizes as soon as it is formed, thereby leaving the underlying metal surface exposed for further attack. This causes rapid and continuous corrosion leading to excessive corrosion. Example, Molybdenum oxide (MoO_3), the oxidation corrosion product of molybdenum is volatile. In closed vapor process (shipping containers), volatile solids such as salts of dicyclohexylamine, cyclohexylamine and hexamethylene amine are used as volatile corrosion inhibitors [52].

11.5.4.6 Synergistic inhibitors

These are single inhibitor which is used in cooling water systems. More often, a combination of inhibitors (anodic and cathodic) is used to obtain better corrosion protection properties [56]. Examples include chromate-phosphates, polyphosphate-silicate, zinc-tannins, and zinc-phosphates.

11.5.4.7 Precipitation inhibitors

These are compounds that forms precipitates on the metal surface, thereby providing a protective film. The most common inhibitors of this category are the silicates and the phosphates. For example, Sodium silicate, is used in many domestic water softeners to prevent the occurrence of rust [57, 58].

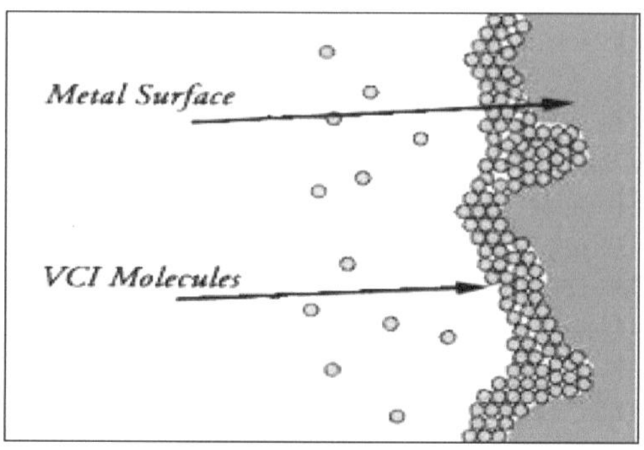

Figure 14.
Schematic representation of vapor phase inhibitors.

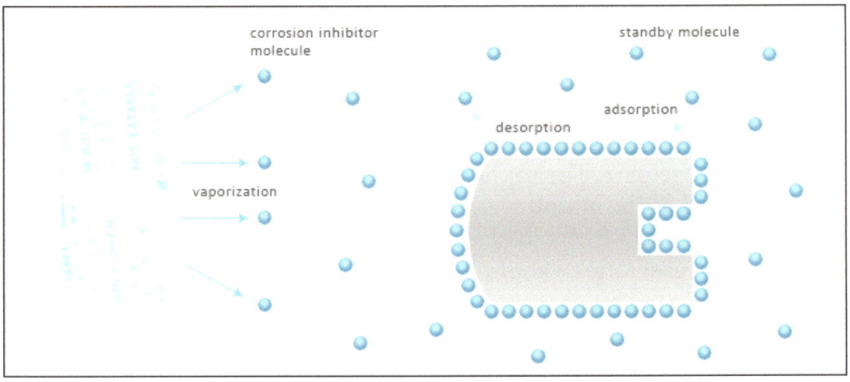

Figure 15.
Schematic representation of volatile inhibitors.

11.5.4.8 Green corrosion inhibitors

There is no clear and accepted definition of "environmentally friendly" or "green" corrosion inhibitors. In practice, corrosion inhibition studies have become oriented towards human health and safety considerations. For this purpose recently, the researchers have been focused on the use of eco-friendly compounds such as plant extracts, expired nontoxic medicines, etc. which contains many organic compounds [59–61]. Amino acids, alkaloids, pigments and tannins are used as green alternatives for the toxic and hazardous compounds. Due to biodegradability, eco-friendliness, low cost and easy availability and the extracts of some common plants and medicinal plant and its products have been studied as corrosion inhibitors for various metals and alloys under different environmental conditions.

11.6 Mechanism of corrosion inhibition

The mechanism of the inhibition process of the corrosion inhibitors under consideration is mainly due to the adsorption. The phenomenon of adsorption is influenced by the nature and surface charge of the metal and by chemical structure of inhibitors. The surface charge of the metal is due to the electrical field which emerges at the interface on immersion in the electrolyte [62–64].

Inhibition usually results from one or more of the following mechanisms

- Adsorption of corrosion inhibitors onto metals

The inhibitive performance is usually depends on the fraction of the surface covered, θ with adsorbed inhibitor. But, at low surface coverage ($\theta < 0.1$), the effectiveness of adsorbed inhibitor species in retarding the corrosion reactions may be greater than at high surface coverage.

- Presence of surface charge on the metal

Adsorption of inhibitor on the metal surface may be due to dipoles of the adsorbed species or electrostatic force of attraction between ionic charges and the electric charge on the metal at the metal/solution interface.

- Effect of functional group and structure

Usually, when the metal contains vacant electron orbitals of low energy such as transition metals. Inhibitors can also bond to metal surfaces by electron transfer to the metal to form a coordinate type of bond. Electron transfer from the adsorbed species is favored by the presence of relatively loosely bound electrons. Example: Anions and neutral organic molecules containing lone pair of electrons or electron systems associated with multiple bonds especially triple bonds or aromatic rings. The electron density at the functional group is directly proportional to the inhibitive efficiency in a series of related compounds.

- Inhibitor and water molecules interaction
  Adsorbed water molecule are removed from the metal surface due to displacement reaction of adsorbed inhibitor molecules and increases the size of hydrocarbon part of inhibitor, which leads to decreasing solubility and increasing adsorption ability. This is consistent with the increasing inhibitive efficiency observed at constant concentrations with increasing molecular size in a series of related compounds.

- Interaction between adsorbed inhibitor species
  Lateral interactions between adsorbed inhibitor species may become significantly increases the surface coverage and the adsorbed species. These interactions either attractive or repulsive. If attractive interactions occur between molecules containing large hydrocarbon components (e.g., n-alkyl chains), may the chain length increases. Then the increasing Van der Waals attractive force between the adjacent molecules leads to stronger adsorption at high coverage.

- Adsorbed inhibitors reaction
  The adsorbed corrosion inhibitor may react usually by chemical or electrochemical reduction to form a product that may exhibit inhibitive action. A process of added small quantity of substance is called as primary inhibition and that due to the reaction product is secondary inhibition. In these cases, the inhibitive efficiency may increase or decrease with time, it depends on the extent of secondary inhibition is more effective than the primary inhibition. For example, sulfoxides can be reduced to sulfides which are more efficient inhibitors.

- Diffusion barrier formation
  The absorbed inhibitor molecules may form a surface layer that acts as a physical barrier to the diffusion of ions or molecules and to or from the metal surface, and hence retard the rate of corrosion reactions. A surface film of these types of inhibitors affects both anodic and cathodic reactions.

- Blocking of reaction sites
  The blocking decreases the number of metal atoms at which corrosion reactions can occur. During this, mechanisms of the reactions are not affected, and the Tafel slopes of the polarization curves remain unaffected [65].

- Electrode reactions
  Corrosion reactions involve the formation of adsorbed intermediate molecules with surface metal atoms. The adsorbed inhibitors will forbid the formation of these adsorbed intermediates, but the electrode processes may proceed by alternative paths through intermediates containing the inhibitor. In this process, the inhibitor act as catalyst and remain unchanged. Such reactions of inhibitor are characterized by an increase in the Tafel slope of the anodic dissolution of the metal. Inhibitors may also retard the rate of hydrogen evolution on the metals by affecting the mechanism of the reaction [66]. This effect has been observed on iron in the presence of inhibitors such as phenylthiourea, aniline derivatives, benzaldehyde derivatives and pyridinium salts [67].

- Electrical double layer alteration

The adsorption of ions or species that can form ions on metal surfaces will change the electrical double layer at the metal/solution interface, and this will affect the rates of the electrochemical reactions [68]. The adsorption of cations such as quaternary ammonium ions and protonated amines makes the potential more positive in the plane of the closest approach to the metal ions from the solution. This positive potential shift hinders the discharge of the positively charged hydrogen ions. These effects have been observed with sulfosalicylate ions and the benzoate ions [69, 70].

Conclusion

- Corrosion is a natural process which reduces the binding energy in metals and degrades the useful properties of materials.

- The end result of corrosion involves a metal atom being oxidized, whereby it loses one or more electrons.

- The corrosion manifests itself as a break-up of bulk metal to metal powder. Corrosion.

- Corrosion inhibitors are a great effective method of preventing corrosion.

- The knowledge of the method of the action, facilitates the choice of the inhibitors, improves efficiency, avoids the process is impaired and side effects.

- It is important in the choice of inhibitor whatever may be the method, ascertain the subsequent effects of this towards the environment.

- The environmental friendly inhibitors have shown excellent results, outperforming conventional inhibitors.

Author details

Geethamani Palanisamy
Department of Chemistry, SNS College of Technology, Coimbatore, Tamil Nadu, India

*Address all correspondence to: drpgeethamani@gmail.com

References

[1] Jones T. International Iron and Steel Institute; steel industry and the environment: Technical and management issues. UNEP/Earthprint. 1997

[2] Boffardi BP. Corrosion inhibitors in the water treatment industry. In: ASM Handbook. Vol. 13A. OH: ASM International, Materials Park; 2003

[3] Holcomb R. Corrosion in supercritical water: Ultrasupercritical environments for power production. In: ASM Handbook. Vol. 13C. OH: ASM International, Materials Park; 2006

[4] Shreir LL, Jarman RA, Burnstein GT. Principles of Corrosion and Oxidation. Vol. 1. 1994

[5] Geethamani P, Kasthuri PK. The inhibitory action of expired asthalin drug on the corrosion of mild steel in acidic media: A comparative study. Journal of the Taiwan Institute of Chemical Engineers. 2016;**63**(4):490-499

[6] Roberge PR. Corrosion Engineering Principles and Practice. New York: McGraw-Hill; 2008

[7] Geethamani P, Kasthuri PK, Aejitha S. Journal of Applied Chemical Science International. 2015;**3**(4):151-157

[8] Geethamani P, Kasthuri PK, Aejitha S. Chemical Science Review and Letters. 2014;**2**(6):507-516

[9] Geethamani P, Kasthuri PK, Aejitha S. Elixir Corrosion & Dye 762014. pp. 28406-28410

[10] Bentiss F, Traisnel M, Lagrenee M. Corrosion Science. 2000;**42**:127

[11] Sharma S, Chaudhary RS. Bulletin of Electrochemistry. 2000;**16**:267

[12] Bentiss F, Lagrenee M, Traisnel M, Hornez JC. Corrosion. 1999;**55**:968

[13] Yurt A, Balaban A, Kandemir SU, Bereket G, Erk B. Materials Chemistry and Physics. 2004;**85**:420

[14] Zaim S, Muralidharan S, Iyer S, Muralidharan B, Vasudevan T. British Corrosion Journal. 1998;**33**:297

[15] Cisse MB, Zerga B, El Kalai F, Touhami ME, Sfaira M, Taleb M, et al. Surface Review and Letters. 2011;**18**:303

[16] Musa AY, Jalgham RTT, Mohamad AB. Corrosion Science. 2012;**56**:176

[17] Nataraja SE, Venkatesha TV, Tandon HC. Corrosion Science. 2012;**60**:214

[18] Thomas JGN. The Mechanism of Corrosion. Oxford, UK: Butterworths Heinemann; 1994

[19] Geethamani, Kasthuri. Physical chemistry. Cogent Chemistry. 2015;**1**:1091558

[20] Ash M, Ash I. Handbook of Corrosion Inhibitors. Texas, Houston, USA: NACE; 2001

[21] Schweitzer PA. Fundamentals of Metallic Corrosion. 2nd ed. London: CRC Press; 2007

[22] Geethamani P, Kasthuri PK, Aejitha S. IJCPS. 2015;**3**(1):1442-1448

[23] Sastri VS. Corrosion Inhibitors: Principles and Applications. Weinheim, Germany: Wiley-VCH; 1998

[24] Mercer AD. Corrosion Inhibition: Principles and Practice. Oxford, UK: Butterworths Heinemann; 1994

[25] Sharma SK. Front Matter, in Green Corrosion Chemistry and Engineering: Opportunities and Challenges.

Weinheim, Germany: Wiley-VCH Verlag GmbH & Co. KGaA; 2011

[26] Aejitha S, Kasthuri PK, Geethamani P. Chemical Science Review and Letters. 2014;**2**(7):566-573

[27] Kelly RG, Scully R, Shoesmith DW, Buchheit RG. Electrochemical Techniques in Corrosion Science and Engineering. New York: Marcel Dekker press; 2002

[28] Bailey JC, Porter FC, Pearson AW, Jarman RA. Aluminum and Aluminum alloys. In: Corrosion: Metal/Environment Reactions. 3rd ed. London: Springer; 1994

[29] Muralidharan S, Iyer S. Journal of the Electrochemical Society of India. 1999;**48**:113

[30] Hariharaputhran R, Subramanian A, Antony AA, Sankar PM, Gopalan A, Vasudevan T, Venkatakrishna S, Iyer S. British Corrosion Journal. 1998;**33**:214

[31] Aejitha S, Kasthuri PK, Geethamani P. Asian Journal of Chemistry. 2016;**28**(2):307-311

[32] Quraishi MA, Rawat J, Ajmal M. Journal of the Electrochemical Society of India. 2000;**49**:35

[33] Lazarova EM, Yankova TN, Neykov GD. Bulgarian Chemical Communications. 1996;**28**:647

[34] Al-Mayouf AM, Al-Suhybani AA, Al-Ameery AK. Desalination. 1998;**116**:25

[35] Touham F, Aouniti A, Abed Y, Hammouti B, Kertit S, Ramdani A. Bulletin of Electrochemistry. 2000;**16**:245

[36] El-Rehim SS, Ibrahim AM, Khaled KF. Journal of Applied Electrochemistry. 1999;**29**:593

[37] Bastidas JM, Polo JL, Cano E, Torres CL. Journal of Materials Science. 2000;**3**:2637

[38] El Rehim SSA, Ibrahim MAM, Khaled KF. Corrosion Prevention & Control. 1999;**46**:157

[39] Du T, Chen J, Cao DZ, Cao C. Bulletin of Electrochemistry. 1997;**13**:13

[40] Ita BI, Offiong OE. Journal of Pure and Applied Science. 1999;**5**:497

[41] Umoren SA, Obot IB, Odewunmi NA. Journal of Dispersion Science and Technology. 2014:922887

[42] Mistry BM, Patel NS, Sahoo S, Jauhari S. Bulletin of Materials Science. 2012;**35**:459

[43] Khaled KF. Electrochimica Acta. 2008;**53**:3484

[44] Umoren SA, Li Y, Wang FH. Corrosion Science. 2010:52, 2422

[45] Aejitha S, Kasthuri PK, Geethamani P. Indian Journal of Applied Research. 2014;**4**(12):218-220

[46] Ebenso EE, Arslan T, Kandemirli F, Love I. International Journal of Quantum Chemistry. 2010;**10**:2614-2636

[47] Sorensen PA, Weinell CE, Dam-Johansen K, Kiil S. Journal of Coating Technology and Research. 2010;**7**:773

[48] Aejitha S, Kasthuri PK, Geethamani. International Journal of Chemical Sciences. 2015;**13**(1):38-52

[49] Aejitha S, Kasthuri PK, Geethamani P. International Journal of Advanced Technology In Engineering Science. 2015;**3**(5):25-30

[50] Abdallah M, AL Jahdaly BA. International Journal of Electrochemical Science. 2015;**10**:9808

[51] Singh AK, Ji G, Prakash R, Ebenso EE, Singh AK. International Journal of Electrochemical Science. 2013;**8**:9442

[52] Singh A, Singh AK, Quraishib MA. The Open Electrochemistry Journal. 2010;**2**:43

[53] Ansari KR, Quraishi MA, Prashant, Ebenso EE. International Journal of Electrochemical Science. 2013

[54] Umoren SA, Obot IB, Gasem ZM. Ionics. 2014:1280

[55] Verma CB, Quraishi MA, Singh A. Journal of the Taiwan Institute of Chemical Engineers. 2015;**49**:229

[56] Fouda AS, Farahat MM, Abdallah M. Research on Chemical Intermediates. 2014;**40**:1249

[57] Hari Kumar S, Karthikeyan S. Journal of Materials and Environmental Science. 2013;**4**:675

[58] Karthik G, Sundaravadivelu M. ISRN Electrochemistry. 2013:10

[59] Megalai SM, Manjula P, Manonmani KN, Kavithaa A, Babyd N. Electrochimica Acta. 2012;**30**:395

[60] Obi-Egbedi NO, Obot IB, Eseola AO. Arabian Journal of Chemistry. 2014;**7**:197

[61] Singh P, Quaraishi MA. Indian Journal of Life Sciences. 2012;**7**:12270

[62] Singh AK, Quraishi MA. Journal of Materials and Environmental Science. 2010;**1**:101

[63] Sudhish KS, Quraishi MA, Ebenso EE. International Journal of Electrochemical Science. 2011;**6**:2912

[64] Junedi S. International Journal of Electroanalytical Chemistry. 2012;**7**:3543

[65] Suraj B, Ade NV, Shitole S, Lonkar M. International Journal of ChemTech Research. 2014;**6**:974

[66] Jain T, Chowdhary R, Arora P. Bulletin of Electrochemistry. 2005;**21**:1

[67] Umoren SA, Obot IB, Obi-Egbedi NO. Journal of Materials Science. 2008;**44**:274

[68] Umoren SA, Solomon MM, Eduok UM, Obot IB, Israel AU. Journal of Environmental Chemical Engineering. 2014;**2**:1048

[69] Vinodkumar KP, Sethuraman MG. Transactions of the SAEST (Society for Advancement of Electrochemical Science and Technology). 1999;**34**:2

[70] Yoshida T, Namba O, Chen L, Okuda T. Chemical & Pharmaceutical Bulletin. 1990;**38**:1113

Chapter 2

Structural Effect in Ionic Liquids Is the Vital Role to Enhance the Corrosion Protection of Metals in Acid Cleaning Process

Perumal Kannan and Anitha Varghese

Abstract

Various kinds of methods have been developed and used to overcome different types of corrosion throughout the world. One possible and easy way to avert corrosion is use of an inhibitor. An inhibitor can be applicable to any type of metal irrespective of medium (acid, alkaline, and neutral). Still, several inhibitors are emerging day by day in the corrosion world and most of them are heterocyclic compounds. In this respect, ionic liquid is attracting the attention of the research community. Because of ionic liquid's salient feature of melting and boiling points, it is being employed as a solvent in various types of reaction. In recent years, synthesizing and functionalizing the structure of ionic liquids in such a way to attain the desire requirement have become significant key factors in the field. By altering the cationic part or anionic part (halogen group), the chemical property of ionic liquids will change considerably. Besides, it will enhance the tendency of the electron-donating nature of the cationic part. This behavior equips them to be employed in the field of corrosion. While it meets the metal surface in the aggressive medium it will be attracted, leads to better surface protection from metal dissolution.

Keywords: ionic liquids, inhibitor, electrochemical analysis, structure effect, surface protection

1. Introduction

Corrosion is defined as the destruction or deterioration of metal because of its reaction with the environment. The age of corrosion is as old as the earth. It is known that it had different names at different time. However, carbon steel is prone to corrosion in an acid environment, which causes material damage and increases downtime costs in industry. Hence, a periodical cleaning and descaling process is performed using inorganic acids to remove the corrosion products. Due to acid aggressiveness, inhibitors were used to reduce metal dissolution [1]. The study of corrosion inhibition and mechanistic processes is a key area of research [2]. A large number of organic compounds having nitrogen, oxygen, sulfur, and phosphorus were synthesized to reduce the corrosion of metal in aggressive medium. The literature also revealed that the organic compounds containing π-bonds, electron density on the heteroatom, planner structure, and aromaticity have supported the effects of adsorption inhibitor molecules on the metal surface.

Over the last few decades, several types of organic compounds have been synthesized and used as corrosion inhibitors to mitigate the corrosion process in different environments [3–5]. More recently, ionic liquids have been recognized as a better corrosion inhibitor for metal corrosion in an acid environment [6–9]. Due to their unique properties such as non-flammability, very low vapor pressure, and boiling and melting point, ionic liquids can be considered to be good corrosion inhibitors. Recently, the software development in the theoretical analysis of molecules to describe plausible corrosion mechanisms has increased considerably [10]. Ionic liquids are salts, a combination of (organic) cation and (organic/inorganic) anion as a liquid form at room temperature; in other words, they are called organic salts, which have melting points below room temperature. The electronic configuration of imidazolium (heterocyclic) depends on the heteroatom of the main carbon skeleton and furthermore causes the interaction between the metal surface and ionic liquid [11]. The significant features of its physical and chemical properties such as liquid state, low vapor pressure, viscosity, and inflammability place ionic liquids in the field of corrosion as inhibitors. The various kinds of cation and anion govern the possibility to have property that required. Many ionic liquids are used as solvents for batteries and organic reactions. The application of ionic liquids in the corrosion field was first reported in 1996. The ionic liquid dissociation pattern gave a clue to the adsorption mode on the metal surface. The adsorption combination of ionic liquids depends on the metal nature and temperature.

The literature shows that inhibition effect mainly depends on alkyl chain length substitution on the N_1/N_2 atoms. A heterocyclic compound containing a longer alkyl chain length would offer better corrosion protection [12]. However, many studies have implied that increasing the carbon chain length on N atoms does not necessarily increase the inhibition effect [13, 14]. Murulana [15] supported the above fact and described that not only does increasing alkyl chain length in the N atom increase protection efficiency but also other atoms close by will enhance the protection effect. The main reason for surface protection is the induction of alkyl substitution on the cationic part of the ionic liquid.

2. Structural impact of ionic liquid in various aspects

2.1 Implication of ionic liquid alkylation in electrochemical studies

Since ionic liquids possess cations and anions, they are adsorbed onto the metal surface anodes and cathodes, respectively. Because of this, metal dissolution considerably decreases. The adsorbing efficiency of the inhibitor purely depends on the electron-donating ability of the ionic liquid active site [16]. Likhanova et al. [17] explained in his research that the electrochemical parameter showed the alkylation of longer chain length on imidazolium and pyridinium and significantly improved charge transfer resistance. Additionally, inhibitor concentration also improved metal dissolution at the metal/solution interface. In the presence of ionic liquid, the anodic and cathodic reactions were affected. $ImDC_{18}$ Br and PyC_{18} Br inhibit the anodic area more than the cathodic area, because of the adsorption between HSO_4^- and Br^-, and consequently, the cathodic reaction is also arrested. The increased polarization resistance indicates the adsorption of ionic liquid on the metal surface to block the active area for the mitigation of corrosion. This efficiency has increased due to the presence of ethylbenzene at the N_3 atom of the imidazole ring. Delocalization of the electron is much favored in the 3-ethylbenzene form of imidazolium ionic liquid than in the acetyl imidazolium form [18]. On the other hand, the (pyridiazinium ionic liquid, S_1) 1-(6-ethoxy-6-oxohexyl) pyridazin-1-ium bromide bears a higher charge transfer resistance value than the (S_2) 1-(2-bromoacetyl) pyridiazinium bromide, which is advocated for reasonable corrosion protection at elevated temperatures. The causes behind better corrosion protection were increased frequency of relaxation [19] as the concentration and temperature increased.

Long alkyl chain length improves the relaxation time at the interface of the metal/ solution interface. Some authors [20] have reported the influence of negative ions of ionic liquids in electrochemical studies. According to Martin et al., corrosion rate decreased with an ionic liquid with the same alkyl chain length but different counter ion (anionic). The surface coverage area obeys the following order: $PF_6^- > BF_4^- > Br^- > Cl^-$. Hence, the ionic radius of the anions is increased by the same amount. The larger anionic radius of PF_6^- turns out to have better protection than the others [14]. The negative ion contribution in the ionic liquid structure could be evaluated through its corrosion protection performance. Saleh et al. discussed the effect of Br^- and Cl^- ions with cetylpyridinium ions in sulfuric acid over mild steel [8, 21].

Corrosion current has decreased considerably compared to the blank medium. The bromide ion has a stronger effect on anodic dissolution than the chloride ion. The anionic radius of bromine atoms is better suited for the bonding and surface coverage area on the metal/solution interface. Likewise, the anionic effect of ethyl sulfate and acetate ions on ethylmethylimidazolium can be deduced by comparing its protection efficiency over mild steel. Among the ionic liquids, [EMIM] [$EtSO_4$] is better at mitigating the anodic curve current than the acetate ion. Most probably, the inhibition efficiency is higher than the acetate ion, and this is attributed to the increased number of electronegative (hetero) atoms in the ionic liquid. On the other hand, the [EMIM] [Ac] ionic liquid showed reduced contribution compared to the [BMIM] [Ac] ionic liquid because of its long alkyl chain length [22].

2.2 Improvement in electrochemical noise resistance due to structural modification of the inhibitor

Electrochemical noise is another technique to ensure corrosion protection of the metal surface offered by ionic liquids. When the metal is exposed to the acid medium, the electrochemical reaction commences as fast as it can. As a result, the current is measured in the form of current deviation (σI). For the blank medium, current noise appears with higher amplitude, vindicating the absence of passivation over the metal surface. Also, the current transient width and amplitude designate the available active surface area on the metal surface. Metal immersed in acid medium afforded the noise resistance value of $1.474 \times 10^4\ \Omega\ cm^2$. These active areas were reduced due to the adsorption of ionic liquid ($[BMMB]^+Br^-$). Thus, the *Rn* value increased from 5.29 to $27.7 \times 10^4\ \Omega\ cm^2$, as the concentration increased from 150 to 250 ppm, respectively. On observing [23], the noise amplitude decreased at each concentration level of ionic liquid. This ionic liquid contributed to corrosion protection by governing its electron-rich center to form a bonding with the metal. On the other hand [24], $[BMEB]^+BF_4^-$ ionic liquid influences the electrochemical noise to reduce gradually that of $[BMMB]^+Br^-$, which is attributed to less electrochemical reaction taking place. As a result, current noise amplitude decreased and concurrently noise resistance increased.

On comparing the noise resistance value of both ionic liquids, it clearly indicates that ethyl-derived ionic liquid has a better tendency to protect the metal surface at the interface than $[BMMB]^+Br^-$. $[BMEB]^+BF_4^-$ possesses higher noise value at each concentration than $[BMMB]^+Br^-$ at the same concentration. Ethyl and BF_4 groups were more susceptible to transferring electrons, which resulted in better corrosion protection in acid medium as shown in **Figure 1**.

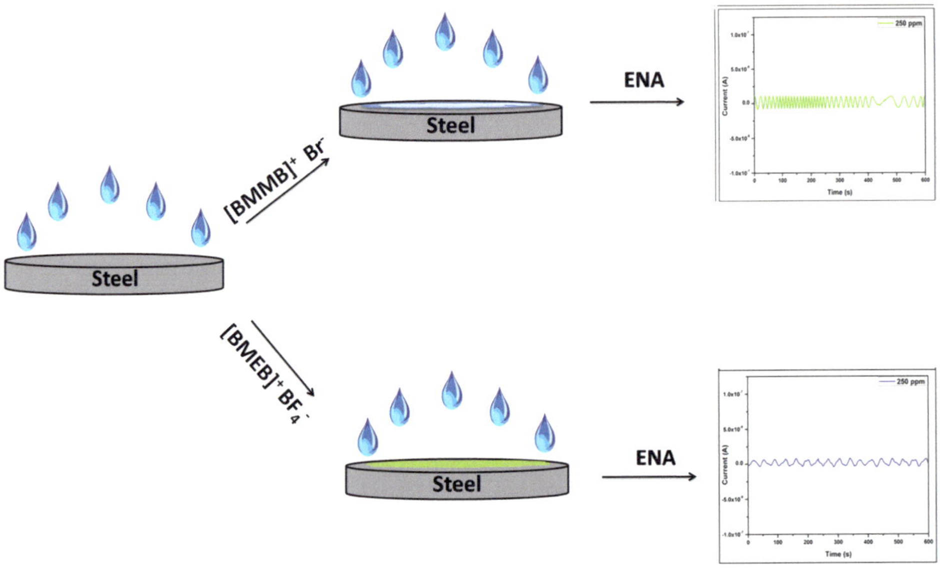

Figure 1.
Schematic representation of electrochemical noise analysis (ENA) of ionic liquids.

2.3 Significance of carbon chain length on surface protection of metal in acid medium

On comparing the surface protection effect of any ionic liquid with its alkylated and non-alkylated form, it would help to understand the structure effect of surface protection. Gabler et al. [25] reported that strong binding between metal surface and ionic liquid is key for corrosion reduction. In an electrochemical study of 2-hydroxy- and butylammonium sulfonyl imide, the C—F bond cleavage was reduced significantly in butyl form compared to the former. On the other hand, Kaczerewska et al. [26] studied the structural effect of inhibition efficiency on metal protection in acid medium. Interconnecting a gemini cation with a bridged oxygen atom showed better surface coverage to avert corrosive ion contact with metal surfaces. The 18-O-18 gemini cationic part of ionic liquid offered improved protection resistance against metal dissolution [27]. Vastag et al. suggested [12] that by increasing the alkyl chain number in N-substituted cations would favor the inhibitor action of organic compounds to isolate metal from further corrosion. Generally, the surface protection of ionic liquid on carbon steel was improved when the N_3 atom was alkylated with carbon chain lengths from n-7 to n-9. Infrared spectra also confirmed the shift in wave number [7]. Image examination of metal surfaces exposed to 1-ethyl [12] and 1-allyl [28] 3-butylimidazolium bromide ionic liquid expressed the inhibition effect of ionic liquid. The inductive effect of the allyl group offered increased electrons in its structure in the cation of ionic liquid. Hence, as resonance increased, the adsorption of ionic liquids over the metal surface also increased. Damage to the negative ions in the ionic liquids influenced corrosion protection [29]. HSO_4^- has greater tendency to render protection on metal surfaces than BF_4^-. Sometimes, iminium compounds reduce surface heterogeneity, caused by adsorption of ionic liquid over the metal surface. Apparently, elemental analysis of the inhibited metal showed that the inhibitor's constituents were 15 wt% of carbon and 10 wt% of oxygen, and also corroborated that dodecyl iminium chloride is more favorably adsorbed than the non-alkylated form [30].

DOI: http://dx.doi.org/10.5772/intechopen.82422

2.4 Study of hydrogen gas evolution over N_1 atom of ionic liquid acid

The hydrogen gas evolution method is one of the ways to determine the corrosion rate of metal in an aggressive medium. Hydrogen gas evolution is a result of a reduction (cathodic) reaction at the metal surface. The volume of hydrogen gas produced in the medium is gradually reduced in the presence of ionic liquid. The increment in alkyl chain length on N_1 atoms of the benzimidazolium cation offers maximum surface coverage on the active site. Thus, a lesser amount of hydrogen gas will be produced than the blank medium [24]. As seen, the volume of hydrogen gas is reduced for both methyl and ethyl benzimidazolium ionic liquid in concentration [23]. Furthermore, it can be understood that each concentration of ethyl benzimidazolium ($[BMEB]^+BF_4^-$) possesses a very low volume of H_2 gas compared to methyl benzimidazolium ($[BMMB]^+Br^-$).

From the above valid point, it is vindicated that the alkyl chain length offers maximum energy or electrons to adhere to the cathode area, which is also the route to sturdy passive film formation between the corrosive medium and the metal surface. The general corrosion mechanism for carbon steel in 1 N HCl acid is described below. As shown in the following equation the ionic liquid is adsorbed onto the anodic and cathodic areas of the carbon steel surface, respectively.

Anodic protection reaction of $[BMMB]^+Br^-$:

$$Fe + (Cl)^-_{ads} \leftrightharpoons (Fe\,Cl)^-_{ads} \quad (1)$$

$$Fe\,(Cl)^-_{ads} + BMMB^+ \leftrightharpoons \left[Fe\,(Cl)^-\,BMMB^+\right]_{ads} \quad (2)$$

$$Fe\,(Br)^-_{ads} + BMMB^+ \rightarrow \left[Fe\,(Br)^-\,BMMB^+\right]_{ads} \quad (3)$$

$[BMEB]^+BF_4^-$:

$$Fe\,(Cl)^-_{ads} + BMEB^+ \leftrightharpoons \left[Fe\,(Cl)^-\,BMEB^+\right]_{ads} \quad (4)$$

$$Fe\,(Br)^-_{ads} + BMEB^+ \rightarrow \left[Fe\,(Br)^-\,BMEB^+\right]_{ads} \quad (5)$$

Likewise in the cathodic area, the reduction (hydrogen gas formation) reaction was considerably decreased due to the alkylation impact of ionic liquid on the adsorption effect instead of H^+, and thus hydrogen gas evolution was reduced.

Cathodic protection reaction of $[BMMB]^+Br^-$:

$$Fe + BMMB^+ \leftrightharpoons Fe\,(BMMB)^+_{ads} \quad (6)$$

$$Fe\,(BMMB)^+_{ads} + e^- \rightarrow Fe\,(BMMB)_{ads} \quad (7)$$

$[BMEB]^+\,BF_4^-$:

$$Fe + BMEB^+ \leftrightharpoons Fe\,(BMEB)^+_{ads} \quad (8)$$

$$Fe\,(BMEB)^{+}_{ads} + e^{-} \rightarrow Fe\,(BMEB)_{ads} \tag{9}$$

As seen in the above equations, the adsorption of the cationic part of ionic liquids is favored in both anodic and cathodic reactive sites. The most favored cationic part among the above three kinds of ionic liquids is $[BMEB]^{+}$. The ($CH_3^{-}\,CH_2^{-}$) ethyl group, which effortlessly offered electrons to the metal surface, explains the reason for the occurrence of reduced volume of hydrogen gas in the acid medium. **Figure 2** shows the hydrogen gas evolution reaction in the blank and benzimidazolium ionic liquid.

2.5 Influence of alkyl chain length on quantum chemical parameters

Quantum chemical studies have been used as an efficient method to evaluate the corrosion inhibition performance of any kind of inhibitor. Since corrosion inhibition is adsorption related, dynamic and quantum studies are used to characterize the electronic properties that help to understand the adsorptive properties of the organic compound [31]. Many factors, including heteroatoms, π-electrons, aromatic rings, and carbon chain length, influence the adsorption properties of the inhibitor with metal surfaces [32]. The adsorption properties of ionic liquids will differ by their structural nature, particularly in HOMO and LUMO energy levels. The higher value of HOMO energy level contributes to the electron offering tendency to the acceptor molecule or lower energy state. Gad et al. [33] reported that increasing alkyl chain length increased HOMO energy level in pyridinium bromide ionic liquid. The maximum HOMO level was attained for (C_{12}) 4-mercapto-1-dodecylpyridinium bromide compared to C_8 and C_{10}.

Likewise, the LUMO energy level of an inhibitor also depends on the functional group present in the structure. The maximum LUMO value is acquired by (C_8) pyridinium bromide ionic liquid compared to the rest. A smaller alkyl chain length will decrease HOMO energy level and increase LUMO energy level. LUMO defines electron-accepting behavior of the molecule from the neighboring environments. Since (C_8) ionic liquid has a higher LUMO level, it can withdraw the electron from the metal, and even from feedback bonding, and result in strong adsorption with the metal surface. The change in energy gap is an important factor for measuring reactivity of ionic liquid to adsorb onto the metallic surface. The lower value of ΔE is the reason for the higher inhibition efficiency in metal corrosion. This is attributed to the

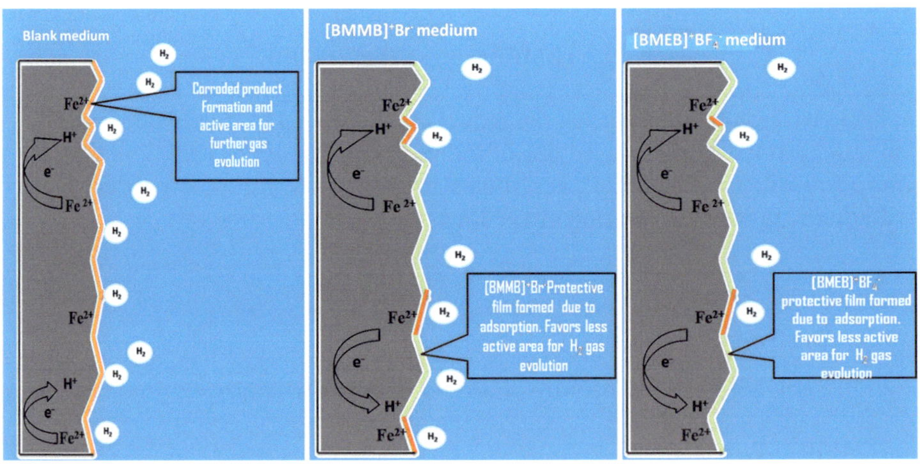

Figure 2.
Pictorial representation of hydrogen gas evolution in the absence and presence of $[BMMB]^{+}Br^{-}$ *and* $[BMEB]^{+}BF4^{-}$ *in 1 N HCl medium.*

minimum amount of energy used to transfer the electron from the highest occupied orbital to the vacant "d" orbital of metal [34]. The ΔE value of the electrodeposition of polyaniline using tetrabutyl phosphonium bromide is less than that using ethyl tributyl phosphonium diethoxy phosphate. The above facts are the same as above in the phosphonium bromide molecule. The energy difference is much higher when substituting the 4-ethoxybenzyl group in the phosphonium [35] cationic part compared to butyl substitution [36]. Likewise, increasing the carbon chain length on the heterocyclic group also decreases the ΔE considerably [37]. The presence of methyl and ethyl groups in the benzimidazolium groups will maintain the minimum energy gap between the HOMO and LUMO groups as represented in **Figure 3**.

In addition, the binding energies of butyl and ethoxy-substituted phosphonium bromide were 2040.9 and 25505.6 kJ, respectively. Ethoxy-substituted phosphonium bromide possessed more negative than butyl-substituted phosphonium bromide. Hence, the ionic liquid is more stable and there is less chance of spitting in the medium because of better passive film on the metal surface. Still, many researchers are studying the structure impact on its corrosion performance. By introducing an alkyl functional group to imidazoline ionic liquid, the relationship between corrosion protection and structure can be discussed. Apparently, partial atomic charges of each atom in the compound describe structural influence using quantum chemical parameters on corrosion protection in the comparison study of [DMIM][BF_4] and [BMIM][BF_4] at corrosion inhibition efficiency; the partial atomic charges focus on cationic moiety rather than anionic moiety, because cations possess large molecular size. Finally, the carbon atom of the alkyl chain contains a negative charge. The C_5 carbon atom has a higher negative charge in [DMIM][BF_4] than in [BMIM][BF_4]. This makes DMIM ionic liquid effective in adsorption on the metal surface against corrosion [38].

In addition, Ibrahim et al. noted their point on structural effect in their research [18]. The adsorption of imidazolium ionic liquid took place through the nitrogen atom of the ring. The coordinate bond occurs between nitrogen and iron. On evaluating two ionic liquids (benzyl and ethyl acetate-substituted imidazolium ionic liquids), benzyl-substituted imidazolium ionic liquid has a higher rate of adsorption on the anodic curve area than ethyl acetate-substituted imidazolium ionic liquid, because the former has a higher rate of relaxation of adsorbed ionic liquid from the metal surface.

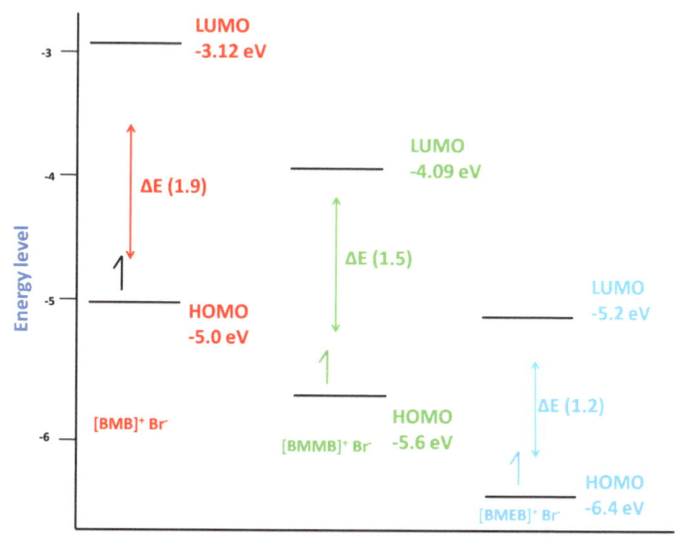

Figure 3.
Schematic representation of energy gap difference between various benzimidazolium ionic liquids.

3. Conclusion

On the basis of the above discussion of the alkylation effect on ionic liquids in various analyses, it can be concluded that structural effect is a considerable key factor in enhancing corrosion protection performance of metal in acid medium. Alkylation of the heterocyclic cation in the ionic liquid, particularly on N_1/N_2 of imidazolium, benzimidazolium, or N_1 of pyridinium or P_1 phosphonium, facilitates the physical/chemical properties of the ionic liquids compared to unalkylated ionic liquid. Not only had the physical/chemical properties changed but also electrochemical properties. This effect has been discussed with respect to various studies, such as electrochemical behavior, surface protection, electronic properties quantum chemical analysis (QCA), and hydrogen gas evolution. Adsorption of ionic liquid is favored mostly when it is being structurally modified as a result of passive film adsorption on the metal surface, which keeps the environment corrosion free. The reason for the extraordinary maturity of the alkylated ionic liquid could be described using theoretical study. This will support the alkylated or structurally modified ionic liquid electronic properties. The electron-donating nature of ionic liquid is being increased by alkylation. As a result, the resonance effect also increased, which contributes to the stability of the ionic liquid compared to the unalkylated form. The use of ionic liquid as a corrosion inhibitor is predominant and more efficient for structural modification.

Acknowledgements

The author is thankful to his management for their help in assisting his work on this chapter. Dr. P. Kannan is acknowledging Centre for Research, CHRIST (Deemed to be University), Bengaluru for carrying out this work.

Conflict of interest

The author of this manuscript has no conflict of interest.

Notes/thanks/other declarations

I express my heartfelt thanks to my friends Mr. P. Mathavan, Dr. L. Mohan, and Dr. K. Vidya for their constant support and encouragement on this endeavor.

Author details

Perumal Kannan* and Anitha Varghese
Department of Chemistry, CHRIST (Deemed to be University), Bengaluru, Karnataka, India

*Address all correspondence to: kannanperumal104@gmail.com

References

[1] Morsi RE, Khamis EA, Al-Sabagh AM. Polyaniline nanotubes: Facile synthesis, electrochemical, quantum chemical characteristics and corrosion inhibition efficiency. Journal of the Taiwan Institute of Chemical Engineers. 2016;**60**:573-581. DOI: 10.1016/j.jtice.2015.10.028

[2] Hegazy MA, Atlam FM. Three novel bolaamphiphiles as corrosion inhibitors for carbon steel in hydrochloric acid: Experimental and computational studies. Journal of Molecular Liquids. 2016;**218**:649-662. DOI: 10.1016/j.molliq.2016.03.008

[3] Farag AA, Ismail AS, Migahed MA. Inhibition of carbon steel corrosion in acidic solution using some newly polyester derivatives. Journal of Molecular Liquids. 2015;**211**:915-923. DOI: 10.1016/j.molliq.2015.08.033

[4] Zarrouk A, Hammouti B, Touzani R, Al-Deyab SS, Zertoubi M, Dafali A, et al. Comparative study of new quinoxaline derivatives towards corrosion of copper in nitric acid. International Journal of Electrochemical Science. 2011;**6**:4939-4952

[5] Wang D, Yang D, Zhang D, Li K, Gao L, Lin T. Electrochemical and DFT studies of quinoline derivatives on corrosion inhibition of AA5052 aluminium alloy in NaCl solution. Applied Surface Science. 2015;**357**:2176-2183. DOI: 10.1016/j.apsusc.2015.09.206

[6] Zhang QB, Hua YX. Corrosion inhibition of aluminum in hydrochloric acid solution by alkylimidazolium ionic liquids. Materials Chemistry and Physics. 2010;**119**:57-64. DOI: 10.1016/j.matchemphys.2009.07.035

[7] Deyab MA, Zaky MT, Nessim MI. Inhibition of acid corrosion of carbon steel using four imidazolium tetrafluoroborates ionic liquids. Journal of Molecular Liquids. 2017;**229**:396-404. DOI: 10.1016/j.molliq.2016.12.092

[8] Atia AA, Saleh MM. Inhibition of acid corrosion of steel using cetylpyridinium chloride. Journal of Applied Electrochemistry. 2003;**33**:171-177. DOI: 10.1023/A:1024083117949

[9] El-Shamy AM, Zakaria K, Abbas MA, El Abedin SZ. Anti-bacterial and anti-corrosion effects of the ionic liquid 1-butyl-1-methylpyrrolidinium trifluoromethylsulfonate. Journal of Molecular Liquids. 2015;**211**:363-369. DOI: 10.1016/j.molliq.2015.07.028

[10] El Belghiti M, Karzazi Y, Dafali A, Hammouti B, Bentiss F, Obot IB, et al. Experimental, quantum chemical and Monte Carlo simulation studies of 3,5-disubstituted-4-amino-1,2,4-triazoles as corrosion inhibitors on mild steel in acidic medium. Journal of Molecular Liquids. 2016;**218**:281-293. DOI: 10.1016/j.molliq.2016.01.076

[11] Zhang QB, Hua YX. Corrosion inhibition of mild steel by alkylimidazolium ionic liquids in hydrochloric acid. Electrochimica Acta. 2009;**54**:1881-1887. DOI: 10.1016/j.electacta.2008.10.025

[12] Vastag G, Shaban A, Vraneš M, Tot A, Belić S, Gadžurić S. Influence of the N-3 alkyl chain length on improving inhibition properties of imidazolium-based ionic liquids on copper corrosion. Journal of Molecular Liquids. 2018;**264**:526-533. DOI: 10.1016/j.molliq.2018.05.086

[13] Guzmán-Lucero D, Olivares-Xometl O, Martínet-Palou R, Likhanova NV, Domínguez-Aguilar MA, Garibay-Febles V. Synthesis of selected vinylimidazolium ionic liquids and their effectiveness as corrosion inhibitors for carbon steel in aqueous sulfuric acid. Industrial & Engineering Chemistry

Research. 2011;**50**:7129-7140. DOI: 10.1021/ie1024744

[14] Yousefi A, Javadian S, Dalir N, Kakemam J, Akbari J. Imidazolium-based ionic liquids as modulators of corrosion inhibition of SDS on mild steel in hydrochloric acid solutions: Experimental and theoretical studies. RSC Advances. 2015;**5**:11697-11713. DOI: 10.1039/C4RA10995C

[15] Murulana LC, Singh AK, Shukla SK, Kabanda MM, Ebenso EE. Experimental and quantum chemical studies of some bis(trifluoromethyl-sulfonyl) imide imidazolium-based ionic liquids as corrosion inhibitors for mild steel in hydrochloric acid solution. Industrial & Engineering Chemistry Research. 2012;**51**:13282-13299. DOI: 10.1021/ie300977d

[16] Kowsari E, Arman SY, Shahini MH, Zandi H, Ehsani A, Naderi R, et al. In situ synthesis, electrochemical and quantum chemical analysis of an amino acid-derived ionic liquid inhibitor for corrosion protection of mild steel in 1M HCl solution. Corrosion Science. 2016;**112**: 73-85. DOI: 10.1016/j.corsci.2016.07.015

[17] Likhanova NV, Dominguez-Aguilar MA, Olivares-Xometl O, Nava-Entzana N, Arce E, Dorantes H. The effect of ionic liquids with imidazolium and pyridinium cations on the corrosion inhibition of mild steel in acidic environment. Corrosion Science. 2010;**52**:2088-2097. DOI: 10.1016/j.corsci.2010.02.030

[18] Ibrahim MAM, Messali M, Moussa Z, Alzahrani AY, Alamry SN, Hammouti B. Corrosion inhibition of carbon steel by imidazolium and pyridinium cations ionic liquids in acidic environment. Portugaliae Electrochimica Acta. 2011;**29**:375-389. DOI: 10.1016/j.corsci.2010.02.030

[19] El-Hajjaji F, Messali M, Aljuhani A, Aouad MR, Hammouti B, Belghiti ME, et al. Pyridazinium-based ionic liquids as novel and green corrosion inhibitors of carbon steel in acid medium: Electrochemical and molecular dynamics simulation studies. Journal of Molecular Liquids. 2018;**249**:997-1008. DOI: 10.1016/j.molliq.2017.11.111

[20] Martins VL, Sanchez-Ramírez N, Calderon JA, Torresi RM. Electrochemistry of copper in ionic liquids with different coordinating properties. Journal of Materials Chemistry A. 2013;**1**:14177-14182. DOI: 10.1039/C3TA12992F

[21] Saleh MM. Inhibition of mild steel corrosion by hexadecylpyridinium bromide in 0.5M H_2SO_4. Materials Chemistry and Physics. 2006;**98**:83-89. DOI: 10.1016/j.matchemphys.2005.08.069

[22] Sasikumar Y, LO O, Bahadur I, Kabanda MM, Obot IB, Ebenso EE. Experimental and theoretical studies on some selected ionic liquids with different cations/anions as corrosion inhibitors for mild steel in acidic medium. Journal of the Taiwan Institute of Chemical Engineers. 2016;**64**:252-268. DOI: 10.1016/j.jtice.2016.04.006

[23] Kannan P, Rao TS, Rajendran N. Anticorrosion behavior of benzimidazolium tetrafluoroborate ionic liquid in acid medium using electrochemical noise technique. Journal of Molecular Liquids. 2016;**222**:586-595. DOI: 10.1016/j.molliq.2016.07.116

[24] Kannan P, Karthikeyan J, Murugan P, Rao TS, Rajendran N. Corrosion inhibition effect of novel methyl benzimidazolium ionic liquid for carbon steel in HCl medium. Journal of Molecular Liquids. 2016;**221**:368-380. DOI: 10.1016/j.molliq.2016.04.130

[25] Gabler C, Tomastik C, Brenner J, Pisarova L, Doerr N, Allmaier G. Corrosion properties of ammonium based ionic liquids evaluated by SEM-EDX,

XPS and ICP-OES. Green Chemistry. 2011;**13**:2869-2877. DOI: 10.1039/C1GC15148G

[26] Kaczerewska O, Leiva-Garcia R, Akid R, Brycki B, Kowalczyk I, Pospieszny T. Effectiveness of O-bridged cationic gemini surfactants as corrosion inhibitors for stainless steel in 3 M HCl: Experimental and theoretical studies. Journal of Molecular Liquids. 2018;**249**:1113-1124. DOI: 10.1016/j.molliq.2017.11.142

[27] Zuriaga-Monroy C, Oviedo-Roa R, Montiel-Sánchez LE, Vega-Paz A, Marín-Cruz J, Martínez-Magadán JM. Theoretical study of the aliphatic-chain length's electronic effect on the corrosion inhibition activity of methylimidazole-based ionic liquids. Industrial & Engineering Chemistry Research. 2016;**55**:3506-3516. DOI: 10.1021/acs.iecr.5b03884

[28] Luo L, Zhang S, Qiang Y, Chen N, Xu S, Chen S. Study on 1-allyl-3-butylimidazalium bromine as corrosion inhibitor for X65 steel in 0.5 M H_2SO_4 solution. International Journal of Electrochemical Science. 2016;**11**:8177-8192. DOI: 10.20964/2016.10.52

[29] Ma Y, Han F, Li Z, Xia C. Acidic-functionalized ionic liquid as corrosion inhibitor for 304 stainless steel in aqueous sulfuric acid. ACS Sustainable Chemistry & Engineering. 2016;**4**:5046-5052. DOI: 10.1021/acssuschemeng.6b01492

[30] Masroor S, Mobin M, Alamb MJ, Ahmad S. The novel iminium surfactant p-benzylidene benzyldodecyl iminium chloride as a corrosion inhibitor for plain carbon steel in 1 M HCl: Electrochemical and DFT evaluation. RSC Advances. 2017;7:23182-23196. DOI: 10.1039/C6RA28426D

[31] Wang HL, Liu RB, Xin J. Inhibiting effects of some mercapto-triazole derivatives on the corrosion of mild steel in 1.0 M HCl medium. Corrosion Science. 2004;**46**:2455-2466. DOI: 10.1016/j.corsci.2004.01.023

[32] Obot IB, Gasem ZM, Umoren SA. Understanding the mechanism of 2- mercaptobenzimidazole adsorption on Fe (110), Cu (111) and Al (111) surfaces: DFT and molecular dynamics simulations approaches. International Journal of Electrochemical Science. 2014;**9**:2367-2378

[33] Gad EAM, Azzam EMS, Halim SA. Theoretical approach for the performance of 4-mercapto-1-alkylpyridin-1-ium bromide as corrosion inhibitors using DFT. Egyptian Journal of Petroleum. 2018;**27**:695-699. DOI: 10.1016/j.ejpe.2017.10.005

[34] Ortaboy S. Electropolymerization of aniline in phosphonium-based ionic liquids and their application as protective films against corrosion. Journal of Applied Polymer Science. 2016;**133**:43923-43935. DOI: 10.1002/app.43923

[35] Kumar S, Goyal M, Vashisht H, Sharma V, Bahadur I, Ebenso EE. Ionic salt (4-ethoxybenzyl)-triphenylphosphonium bromide as a green corrosion inhibitor on mild steel in acidic medium: Experimental and theoretical evaluation. RSC Advances. 2017;7:31907-31920. DOI: 10.1039/C6RA27526E

[36] Bhrara K, Kim H, Singh G. Inhibiting effects of butyl triphenyl phosphonium bromide on corrosion of mild steel in 0.5 M sulphuric acid solution and its adsorption characteristics. Corrosion Science. 2008;**50**:2747-2754. DOI: 10.1016/j.corsci.2008.06.054

[37] Kannan P. Synthesis, characterization and corrosion protection performance of benzimidazolium ionic liquids for

carbon steel in acid environment [thesis]. Chennai: Anna University; 2017

[38] Manamela KM, Murulana LC, Kabanda MM, Ebenso EE. Adsorptive and DFT studies of some imidazolium based ionic liquids as corrosion inhibitors for zinc in acidic medium. International Journal of Electrochemical Science. 2014;**9**:3029-3046

Chapter 3

Exploring *Musa paradisiaca* Peel Extract as a Green Corrosion Inhibitor for Mild Steel Using Factorial Design Method

Olusola S. Amodu, Moradeyo O. Odunlami, Joseph T. Akintola, Seteno K. Ntwampe and Seide M. Akoro

Abstract

The suitability of *Musa paradisiaca* (banana) peel extract as a green corrosion inhibitor for mild steel in acidic medium (1 M HCl) was investigated using factorial method of the design of experiment. The effects of two independent variables (concentration of banana peel extract and temperature) on the corrosion inhibition efficiency were investigated. The physicochemical properties of the extract such as surface tension, viscosity, flash point, and specific gravity were determined using standardized methods provided by the American System of Testing Materials (D-971). The relationship between the independent variables and the inhibitor efficiency was modeled by gasometric and thermometric methods. The statistical analysis of the inhibition efficiency was carried out using the "Fit Regression Model" of Minitab® 17.0, while the fitness of the models was assessed by the coefficient of determination (R^2) and the analysis of variance (ANOVA). From the results obtained, gasometric method achieved a maximum inhibition efficiency of 66.83%, with an R^2 of 90.76%, whereas thermometric method gave a maximum inhibition efficiency of 65.70%, with an R^2 of 95.56%. This study shows that banana peel extract has the capacity to prevent the corrosion of mild steel in acidic medium.

Keywords: banana peel extract, biomass, corrosion, inhibitors, factorial design

1. Introduction

In 2016, NACE International estimated the global cost of corrosion at US$ 2.5 trillion annually. This accounts for about 3.4% of the global gross domestic product (GDP). In the same study, it was discovered that if corrosion prevention best practices are implemented, there could be global savings of between 15 and 35% of the cost of damage [1]. In spite of the technological advancement of this generation, high profile cases of corrosion have continued to emerge [2–4]. Moreover, the extensive application of acid solutions in industrial cleaning and descaling of mild steel makes metal dissolution a common phenomenon Gadiyar et al. [5]. In order to prolong the lifespan of mild steel, to enhance its viability, and to reduce the high cost of production, practical steps need to be employed in corrosion prevention. Failure to prevent or manage corrosion can result to metal losses, loss of production time, leaking vessels, and unwarranted cleanup costs.

Corrosion is an electrochemical process; it is the propensity for metals to revert to their natural ore state. It takes place in the presence of moisture and oxygen, involving chemical reaction and the flow of electrons on the surface of the corroded cells, which greatly accelerate the transformation of metal back to the low-grade ore. The process involves the oxidation of a metal atom, whereby it loses one or more electrons. The resultant effect of corrosion is metal degradation, that is, the breakup of bulk metal, causing it to lose its useful properties [6]. This electrochemical process, often referred to as galvanic cell, occurs when two different metals in physical or electrical contact are immersed in a common electrolyte with different concentrations. Consequently, the more active metal (anode) gets corroded while the more noble metal (cathode) is protected [7, 8]. The fundamental chemical reactions occurring at the anode and cathode are:

$$\text{Anode surface:} F_e \rightarrow F_e^{\,2+} + 2e^-$$

$$\text{Cathode surface:} \frac{1}{2}O_2 + H_2O + 2e^- \rightarrow 2OH^-$$

$$\text{At anode:} F_e^{\,2+} + 2OH^- \rightarrow F_eO.(H_2O)\text{–hydrated iron oxide (brown rust)}$$

Galvanic corrosion is the most common type of corrosion and it occurs regularly in marine vessels, metal structures, and oil pipelines. Furthermore, the phenomenon is commonly observed in water treatment plant, boilers, storage vessels, oil pipelines, etc. In fact, what makes corrosion challenging is that it starts on the internal part of the metal structure, which makes early detection difficult. Other than water and oxygen, galvanic corrosion is affected by: types of metal, agitation, the presence and type of inhibitors, and environmental factors (pH, temperature, humidity, salinity, etc.) [9]. In addition, the dissolution of mild steel in HCl is given as:

$$2F_e + H^+ \rightarrow F_e^{\,2+} + H_2$$

The rate of this reaction is dependent on: metal (its position in the electromotive series), acidity, ferrous ion concentration (by the law of mass action, the increase in ferrous ions should correspond to the decrease in rate of corrosion), and hydrogen gas evolution.

Three techniques are often used for the assessment of corrosion rates namely, weight loss technique, electrochemical impedance spectroscopy, and hydrogen gas evolution method. The weight loss method is considered the most fundamental, against which the accuracy of the other methods is determined. However, the limitations of this method are: (1) the weight loss expressed is the average of the weight of the corroding specimen over a period of time but the changes in corrosion rate over this period is not accounted for; (2) in order to accurately determine the weight loss caused by corrosion, all the corroded particles need to be removed from the specimen surface without removing the uncorroded metal, which practically is unrealistic. Electrochemical technique has been successfully used in many corrosion studies to determine the rate of corrosion [4, 7, 10]. Particularly, it has certain advantages over the weight loss method, due to its ease of corrosion rates determination. With this method, instantaneous corrosion rates as well as changes in corrosion rates over a period of time can be determined. However, during the electrochemical dissolution process for some metals and metal alloys, the atypical polarization performance at the anode is a challenge. Moreover, hydrogen removal can occur in two ways: hydrogen gas evolution and depolarization via oxidation by dissolved oxygen or by some other oxidizing agent [11]. Hydrogen gas evolution method seems to be the most significant and reliable in assessing the rate of corrosion as the mole of metal dissolved directly correlates to the amount of hydrogen gas given off [12].

The application of inhibitors is one of the practical ways to protect metals against corrosion, especially in acidic media. Basically, they function by serving as integuments on metal surface thereby preventing it from chloride ions and oxygen dissolution. Corrosion inhibitors find application in minimizing metallic waste in engineering materials, in addition to the advantages of versatility and cost effectiveness when compared to other corrosion protection methods [10, 13]. Most of the effective inhibitors are either from biomass precursors or chemical compounds containing hetero-atoms such as oxygen, nitrogen, and sulfur with multiple bonds in their molecules through which they are adsorbed on the metal surface [14–16]. This adsorption depends on certain physiochemical properties of the inhibitors: functional group, electron density at the donor atom, n-orbital character, and the electron structure of the molecule. They contain electronegative functional groups and π-electrons in their double bonds, which facilitate their adsorption onto the metal surface. Hence, the strength of an inhibitor to either prevent corrosion reaction from being initiated or slow down the rate of corrosion, is dependent upon the molecular structure of the inhibitor molecules.

The major concern with most chemical inhibitors is their toxicity to the environment. Although many of these synthetic compounds have shown good anti-corrosive activity; their applications have been limited due to environmental considerations [17–20]. Particularly, inorganic corrosion inhibitors such as lead and chromium have been found to constitute significant health challenge to human when released into the environment. This has necessitated the quest for environmentally benign precursors as corrosion inhibitors, for which a plethora of organic materials have been reported. Some of these materials are plant extracts of kola nut, tobacco, *Rosmarinus officinalis*, *Cassia auriculata*, *Argemone Mexicana*, unripe fruit of *Musa acuminate*, roasted coffee seed (*Coffea arabica*), *Carica papaya* leaves, cannabis plant, pomegranate, etc. [14, 19, 21–23]. Furthermore, research has revealed that the basic components of plant extracts are alkaloids, steroids, sugars, gallic acid, tannic acid, and flavonoids, which have been found to improve the protection of metal surface against corrosion [24–26].

The use of natural organic plant extracts as corrosion inhibitors is sometimes referred to as green corrosion prevention. These natural organic compounds are relatively cheaper, nontoxic, and readily available, either as agro-waste and/or agro-industrial waste [27]. Therefore, the aim of this study was to investigate the effectiveness of *Musa paradisiaca* (banana) peel extract as a green corrosion inhibitor for mild steel in acidic medium. The rate of corrosion was assessed using the hydrogen gas evolution method. In addition, the interaction effects of concentration and temperature variation on the corrosion inhibition were assessed using full factorial design and analyzed with relevant statistical tools.

2. Materials and methods

2.1 Preparation of samples and corrosion test solutions

Banana peels were sourced locally. The peels were washed under running water and air dried until a constant mass was recorded. It was milled into powder using a hammer-mill and ball-mill to achieve finesse of powder (about 0.25 mm diameter size). The powdered peel was extracted using 95% ethanol. Five grams of powder

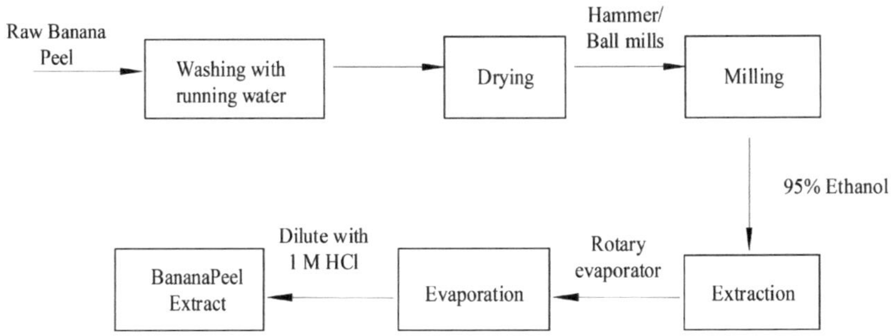

Figure 1.
Flow process for banana peel extract (BPE) as corrosion inhibitor.

was dissolved in 200 ml ethanol for 14 days and thereafter filtered. The filtrate was rotatory evaporated in order to remove excess ethanol, and then diluted with 1 M HCl in distilled water to obtain the corrosion inhibition test solutions in the concentration ratio of 1.0, 2.5, 5, 7.5, and 10% (v/v). **Figure 1** shows the various stages involved in the extraction of banana peels.

In addition, the mild steel used was mechanically press-cut into coupons of dimensions 4 × 2.5 × 0.1 cm. Each coupon was degreased by washing with ethanol, dried in acetone, and immediately transferred into the simulated test solutions. Note that the dried coupons can be preserved in a desiccator until use. Similarly, control experiments were set up but without the addition of the inhibitor. All reagents used were of analytical grade. Banana peel is composed of starch (3%), total dietary fiber (43.2–49.7%), crude fat (3.8–11%), crude protein (6–9%), polyunsaturated fatty acids, pectin, micronutrients (K, P, Ca, and Mg), and amino acid [28]. Also, the mild steel sheet used has the following compositions (% wt): Fe—99.3, Mn—0.34, Cu—0.069, Co—0.069, Ca—0.087, Ni—0.043 and Al—0.03.

2.2 Evaluating the physical properties of banana peel extract as a corrosion inhibitor

2.2.1 Viscosity measurement

This was determined by the Cannon-Fenske viscometer and a circulatory bath with temperature control. Viscosity was calculated using ASTM Method D445–97 [29]. The viscosity η of each sample was calculated using the formula below:

$$\eta = k\rho T, \tag{1}$$

where k is the instrument constant, ρ is the density of banana peel extract sample, and T efflux time (sec) for banana peel extract sample.

2.2.2 Specific gravity determination

The extract of banana peel was transferred into a narrow glass cylinder (SP0121-V Osaka, Japan) and a hydrometer was set into the sample and allowed to stabilize. The value of the specific gravity was taken from the markings on the stem of the hydrometer at the surface of the extract sample.

2.2.3 Surface tension determination

This study employed the American System of Testing Materials D-971 [29] method. Two grams of banana peel extract was added to 50 ml of distilled water in a 100 ml beaker. A platinum ring was then lowered into the solution of banana peel extract in the beaker. It was then brought up to the water sample interface, where the actual measurement takes place. The force required to pull the ring through the interface was measured by a tension meter as the surface tension of the extract solution (dynes cm^{-1}).

2.2.4 Flash point measurement

The measurement of the flash point for the BPE sample was done using ASTM D-92 method [29]. An open cup containing BPE sample was heated at a specific rate while flame was periodically passed over its surface. The lowest temperature at which the BPE vapor ignites without sustaining the flame was recorded as the flash point.

2.3 Evaluating the corrosion inhibition efficiency of banana peel extract

2.3.1 Gasometric method

This method was adopted in this study as described by Ekpe et al. [23], and carried out at the following temperatures: 303, 308, 313, 318, and 323 K, which were achieved using a water bath. The coupons immersed in the prepared test solutions were recovered after 6 h, washed in detergent solution, and rinsed with distilled water, and air dried. The volume of gas evolved from the cathodic reaction during the corrosion process was determined. Hence, gasometric method correlates the quantity of gas evolved to the rate of corrosion. The graph of the volume of gas liberated per minute gives the rate of gas evolution, while the inhibition efficiency ($\mathcal{E}$) and degree of surface coverage (θ) were determined from Eqs. 2 and 3, respectively.

$$\mathcal{E} = \left(1 - \frac{V^{*}_{H}}{V^{0}_{H}}\right) \times 100 \tag{2}$$

$$\theta = \left(1 - \frac{V^{*}_{H}}{V^{0}_{H}}\right) \tag{3}$$

where V^{*}_{H} is volume of hydrogen gas evolved at time t in the presence of inhibitor and V^{0}_{H} is the volume of hydrogen evolved in the absence of inhibitor.

2.3.2 Thermometric method

Temperature determination was carried out as reported by Ebenso et al. [30]. Using the value for the rise in temperature per minute, the reaction number (RN) was calculated as shown in Eq. 4:

$$RN\left({}^{\circ}C/min\right) = (Tm - Ti)/t \tag{4}$$

where T_m and T_i are the maximum and initial temperatures, respectively, attained by the system and t is the time. Similarly, the inhibition efficiency was determined by the reaction number correlation (Eq. 5).

$$\mathcal{E} = \left(\frac{RN_o - RN_i}{RN_o}\right) \times 100 \tag{5}$$

where RN_o is the reaction number of solution without inhibitor, while RN_i is the reaction number of solution with inhibitor.

2.4 Effects of concentration and temperature variations on corrosion inhibition

The data obtained from the banana peel extraction experiments were analyzed statistically using factorial method to obtain a linear fit for the inhibition of mild steel corrosion under varied concentration and temperature. The mathematical model was generated by a MINITAB 17.0. Regression analysis was performed to correlate the response variable to the independent variables. The quality of the fit of the model was evaluated using analysis of variance (ANOVA), where the response was the inhibition efficiency; while concentration and temperature of BPE solution were the input variables.

3. Results and discussion

3.1 Physical properties of BPE

The physical properties determined for the banana peel extract are presented in **Table 1**.

The viscosity of the banana peel extract (BPE) compares favorably well with those previously reported for BPE and other biomaterials. For instance, similar viscosities were reported for extracts from *Citrullus lanatus*, *Phyllanthus*, and banana peels [31, 32]. Viscosity is the property of a fluid that makes it resist flow and sustain frictional force. It is an important property of a good inhibitor, which makes it stick to the surface of metals thereby forming a protective barrier against corrosion. It represents the ability of the extract to adhere longer to metal surfaces, thus enhancing corrosion inhibition. Corrosion inhibitors often contain one or more surfactants, which lowers the surface tension of corrosive fluids [33]. Similarly, the flash point of BPE extract is within the range reported by Kliskic et al. [34] and Betiku et al. [35]. This indicates the flammability or combustibility of the inhibitor.

3.2 Evaluation of BPE efficiency as a corrosion inhibitor

3.2.1 Gasometric method

The efficiency of corrosion inhibition by BPE was determined using gasometric method according to Eq. 1, for the various concentrations of 1.0, 2.5, 5.0, 7.5, and 10.0 g/L (**Table 2**).

Density (g/L cm^3)	Dynamic viscosity (cp)	Surface tension (dynes cm^{-1})	Specific gravity	Flash point (°C)
1.56	32.04	18.0	1.55	237

Table 1.
Physical properties of banana peel extract (BPE).

Levels	1	2	3	4	5
Concentration (g/L), X_1	1.0	2.5	5.0	7.5	10.0
Temperature (K), X_2	303	308	313	318	323

Table 2.
Experimental design.

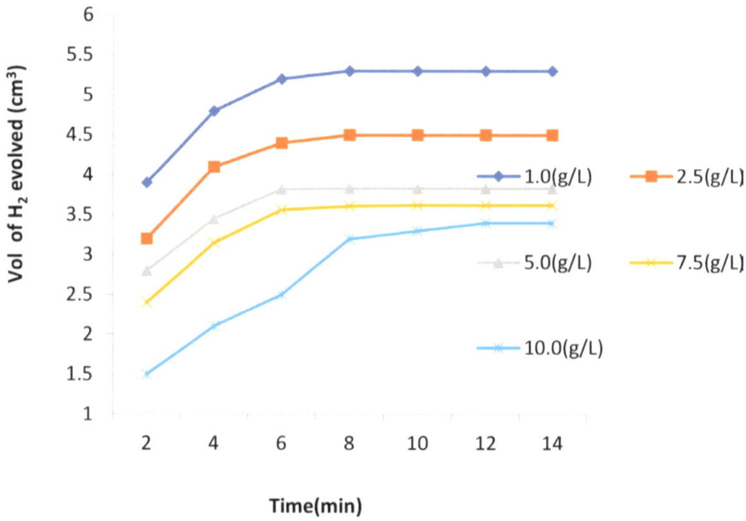

Figure 2.
Volumetric rate of hydrogen gas during corrosion of mild steel in 1 M HCl acid solution.

The volume of gas evolved from the cathodic reaction during the corrosion study (**Figure 2**) is correlated to inhibition efficiency by gasometric method.

As shown in **Figure 2**, the volume of H_2 evolved decreased with increasing concentration of the extract. This can be attributed to increased adsorption forces at higher concentrations. However, hydrogen gas evolution increased with time, until after 8 min before equilibration. This is expected since adsorption decreases with time. From the chemical equation, the dissolution of 2 M of ferrous generates one molecule of hydrogen gas. This means that the evolution of one molecule of hydrogen gas corresponds to the dissolution of 2 M of iron. Theoretically, by mole ratio, it implies that the rate of iron dissolution doubles the rate of hydrogen gas evolution. From the anodic and cathodic reactions, the flow of chloride ion from cathode caused the dissolution of iron at the anode and the metal gets corroded, which is the formation of hydrated iron (II) chloride. Hence, prevention of corrosion is possible if the chloride ion is prevented from having contact with the metal surface. The corrosion of mild steel in HCl solutions has been reported to be a first order reaction [36], which increases with increased acid concentration. In addition, the dissolution of iron steel in hydrochloric acid is dependent upon chloride ion over acidic range of pH. Also, the pH of HCl solution decreased with increased immersion time of the mild steel (**Figure 3**). This was obviously due to increased acidity of the solution.

Moreover, as described in Section 2.3, the volume of hydrogen gas evolution was used to determine the corrosion inhibition efficiency at different temperatures and concentrations of the biomass extract. Consequently, experimental design was used to assess the effects of these independent parameters that ultimately led to peak process performance and the discovery of optimum conditions. The experimental design was generated using a MINITAB 17.0 software (Stat-Ease Inc., USA). Each variable was analyzed at five levels with a total of 25 experiments being performed representing a full factorial. The system response, which is the corrosion inhibition efficiency, was determined by gasometric and thermometric methods (**Tables 3** and **4**).

3.2.2 Thermometric method

Similarly, using thermometric method given in Eq. 4, the percentage corrosion inhibition was evaluated for the concentrations 1.0, 2.5, 5.0, 7.5, and 10.0 g/L, and

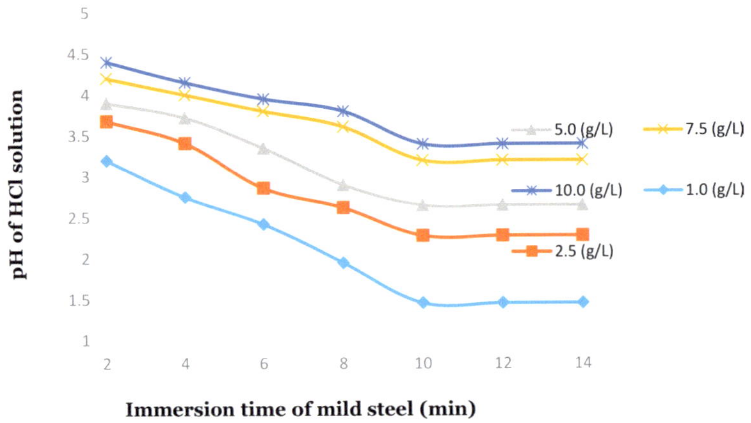

Figure 3.
Variation of pH of HCl solution with immersion time of mild steel in the presence of banana peel extract (BPE).

Run order	Concentration (g/L) (X_1)	Temperature (K) (X_2)	Inhibition, Ɛ (%)
1	7.5	303	64.14
2	5.0	313	51.11
3	1.0	303	44.40
4	10.0	313	61.75
5	2.5	308	49.53
6	7.5	318	49.25
7	7.5	313	53.14
8	2.5	318	40.28
9	5.0	308	56.31
10	2.5	313	44.43
11	10.0	323	49.21
12	1.0	323	24.75
13	2.5	303	54.73
14	10.0	303	72.03
15	1.0	313	34.00
16	7.5	308	59.00
17	7.5	323	44.17
18	1.0	318	29.15
19	5.0	323	41.62
20	5.0	303	61.51
21	1.0	308	39.20
22	5.0	318	29.15
23	10.0	308	66.83
24	10.0	318	54.62
25	2.5	323	34.92

Table 3.
Corrosion inhibition efficiency determination using gasometric method.

Run order	Concentration (g/L) (X_1)	Temperature (K) (X_2)	Inhibition, Ɛ (%)
1	7.5	303	62.4
2	5.0	313	41.93
3	1.0	303	41.58
4	10.0	313	59.85
5	2.5	308	49.25
6	7.5	318	44.98
7	7.5	313	51.73
8	2.5	318	33.94
9	5.0	308	54.13
10	2.5	313	40.31
11	10.0	323	48.77
12	1.0	323	21.75
13	2.5	303	51.75
14	10.0	303	71.56
15	1.0	313	29.95
16	7.5	308	57.92
17	7.5	323	40.86
18	1.0	318	24.86
19	5.0	323	37.72
20	5.0	303	60.01
21	1.0	308	35.32
22	5.0	318	41.93
23	10.0	308	65.70
24	10.0	318	53.62
25	2.5	323	31.56

Table 4.
Corrosion inhibition efficiency determination using thermometric method.

at various temperatures. The experimental results presented in **Tables 3** and **4** show that at constant concentration of the extract; say, at 10 or 7.5 g/L, corrosion inhibition efficiency decreased with increasing temperature. In the same vein, it was observed that at constant temperature; say, 303 K, corrosion inhibition efficiency increased with increasing inhibitor concentration. In a similar study, Gunavathy and Murugavel [37] have reported that the inhibition efficiency of mild steel corrosion in acid medium by *Musa acuminata* fruit peel extract increases with the increase in concentration but decreases with increase in temperature. The same trend was reported by Mayanglambam et al. [38] for *Musa Paradisiaca* extract on mild steel in sulfuric acid solution. Lai et al. [39] have also worked on the inhibition of mild steel in HCl acid solution with a synthetic inhibitor, and/or synthetic mixed-type inhibitors as revealed by potentiodynamic polarization measurement. It was equally reported that the efficiency of the inhibitors decreased with increasing temperature as well as acid concentration.

3.3 Modeling and statistical analysis

Suitable statistical models were chosen to model the interactions between the different experimental variables and their effect on the efficiency of corrosion inhibition, based on the “Fit Regression Model” of MINITAB 17.0 (Pen, USA). The response was modeled with a response surface quadratic model and further analyzed by analysis of variance (ANOVA) to assess the significance of each variable on corrosion inhibition. An empirical model that could relate the response measured to the independent variables was obtained using multiple regression analysis. The response (Y), can be represented by the following quadratic model:

$$Y = \alpha_o + \sum_{i=1}^{n} \alpha_i X_i + \sum_{i=1}^{n} \alpha_{ii} X^2_{\ i} + \sum_{i=1}^{n-1} \sum_{j=i+1}^{n} \alpha_{ij} X_i X_j + \varepsilon \tag{6}$$

where $X_1, X_2, X_3, \ldots, X_n$ are the independent coded variables, α_0 is the offset term, and α_i, α_{ii}, and α_{ij} account for the linear, squared, and interaction effects, respectively, and ε is the random error. A model reduction may be expedient, if there are many redundant model terms [40].

The statistical model summary based on the Lack-of-Fit Test explained the fitness of quadratic models. Using ANOVA to assess the significance of each variable in the model, empirical quadratic models were obtained from Eq. 6. These models, Eqs. (7) and (8), for gasometric and thermometric methods, respectively, were used to predict the efficiencies of the corrosion inhibition at the various values of the independent variables.

$$Y = 376.0 + 2.660\ X_1 - 1.0910\ X_2. \tag{7}$$

$$Y = 376.4 + 2.928\ X_1 - 1.1038\ X_2. \tag{8}$$

The ANOVA of the quadratic regression model for the corrosion inhibition showed the significant level of the model at 90.72 and 95.56% for gasometric and thermometric methods, respectively (**Tables 5** and **6**). It indicates how well the model fits the experimental data, implying that the total variance in the response could be explained using this model. The closeness in the values of R-sq (adj) and R-sq (pred) in both methods also shows the significance of the model.

Source	DF	Adj SS	Adj MS	F-value	P-value
Regression	2	3372.9	1686.45	107.59	0.000
Conc. of BPE (g/L)	1	1885.6	1885.59	120.30	0.000
Temperature (K)	1	1487.3	1487.31	94.89	0.000
Error	22	344.8	15.67		
Total	24	3717.7			
Model summary					
S	R-sq	R-sq (adj)	R-sq (pred)		
3.95907	90.72%	89.88%			
Coefficient					
Terms	Coef	SE coef	T-value	P-value	VIF
Constant	376.0	35.1	10.72	0.000	
Conc. of BPE (g/L)	2.660	0.243	10.97	0.000	1.00
Temperature (K)	−1.091	0.112	−9.74	0.000	1.00

Table 5.
ANOVA for corrosion inhibition efficiency of BPE (gasometric method).

Source	DF	Adj SS	Adj MS	F-value	P-value
Regression	2	3807.7	1903.87	236.50	0.000
Conc. of BPE (g/L)	1	2284.8	2284.78	283.81	0.000
Temperature (K)	1	1523.0	1522.97	189.18	0.000
Error	22	177.1	8.05		
Total	24	3984.9			
Model summary					
S	R-sq	R-sq (adj)	R-sq (pred)		
2.83731	95.56%	95.15%	94.36%		
Coefficient					
Terms	Coef	SE coef	T-value	P-value	VIF
Constant	376.0	25.1	14.97	0.000	
Conc. of BPE (g/L)	2.928	0.174	16.85	0.000	1.00
Temperature (K)	−1.1038	0.0803	−13.75	0.000	1.00

Table 6.
ANOVA for corrosion inhibition efficiency of BPE (thermometric method).

Regression results often show the statistical correlation and importance between the predictor and response. The coefficient of determination (R^2) is the percentage of inhibition efficiency (%) variation that is explained by its relationship with concentration (g/L) and temperature (K). Therefore, the adjusted R^2 is the percentage of inhibition efficiency (%) variation that is explained by its relationship with concentration (g/L) and temperature (K), adjusted for the number of predictors in the model. This adjustment is important because R^2 for this model increases when a new independent variable is added. The adjusted R^2 is a useful tool for comparing the explanatory power of models with different numbers of predictors. P-value for each coefficient tests the null hypothesis that the coefficient has no effect [40].

3.4 Graphical representation of the model

Graphical representation allows easy interpretation of experimental results and the prediction of optimal conditions. From the contour plots, **Figures 4** and 5, BPE was most efficient at temperatures below 307.5 K and concentrations above 8 g/L. The least inhibitory effect was observed at temperature of 318 K and at lower concentrations. This corroborated the experimental results where the highest corrosion inhibition was 72.03 and 71.56%, for gasometric and thermometric methods, respectively. This peak performance occurred when concentration was 10 mol/L and temperature 303 K. Other higher values determined for inhibition efficiency occurred in the neighborhood of 303 and 308 K, and 10 mol/L. Furthermore, the interactive effect of the concentration and temperature on the system's response (inhibition efficiency) was assessed by plotting three-dimensional curves of the response against the independent variables (**Figures 6** and 7). The response distribution in this experiment with respect to the variation of the independent variables shows that temperature has a greater effect.

Similar corrosion inhibition efficiencies have been reported for biomass extracts [37, 41]. In some other studies, inhibition efficiencies in the range of

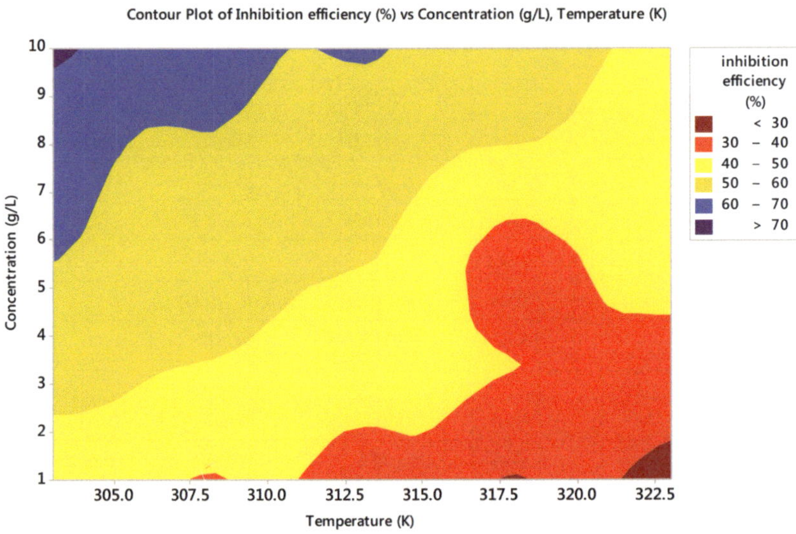

Figure 4.
Contour plot showing the effects of concentration and temperature on the efficiency of corrosion inhibition of BPE (Gasometric method).

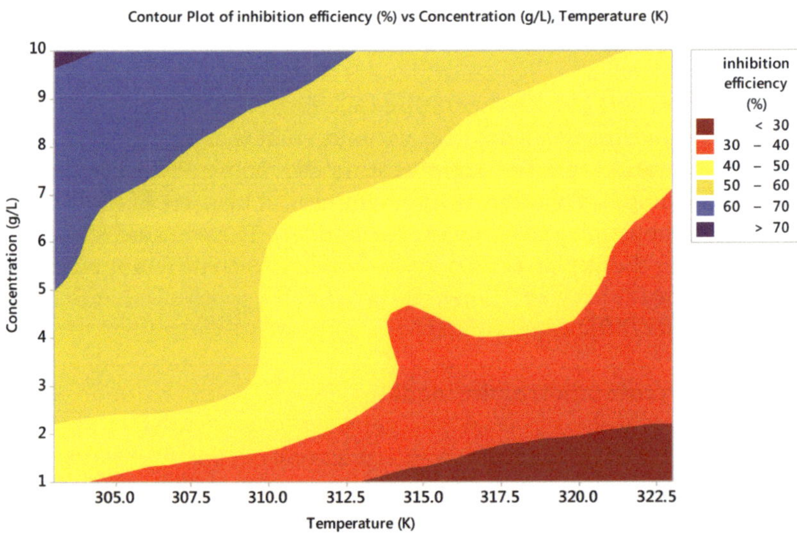

Figure 5.
Contour plot showing the effects of concentration and temperature on the efficiency of corrosion inhibition of BPE (Thermometric method).

80–90 have been reported for mild steel in HCl solution [42–45]. In a study carried out by Ong and Karim [46], where the extract of red onion was used to inhibit corrosion of mild steel in HCl solution, an inhibition efficiency of 90% was reported also at temperature of 303 K. Since a combination of factors such as temperature, concentration of inhibitors, and immersion time affects inhibition efficiency, it is pretty difficult to compare extracts of different biomass. However, reports have shown that temperature and concentration of inhibitors are the predominant factors [47].

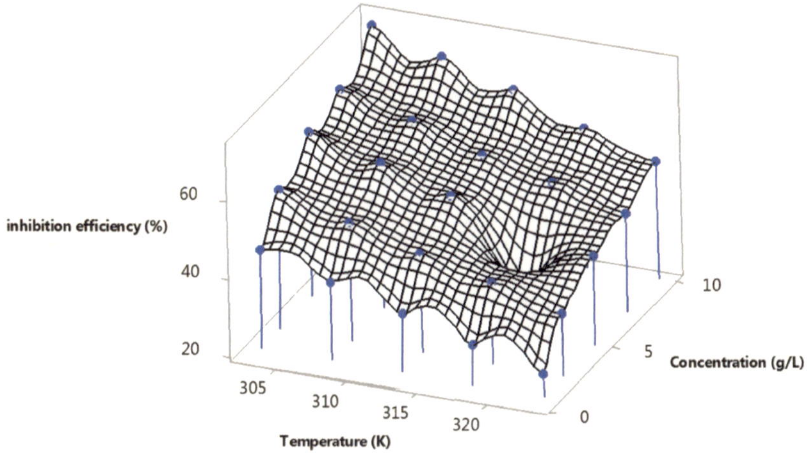

Figure 6.
Surface plot of inhibition efficiency (%) against concentration (g/L) and temperature (K) for gasometric method.

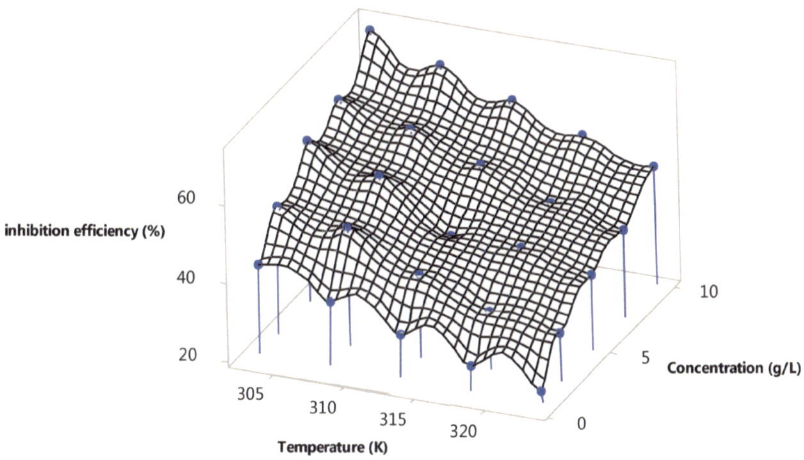

Figure 7.
Surface plot of inhibition efficiency (%) against concentration (g/L) and temperature (K) for thermometric method.

4. Conclusion

Statistical analysis using full factorial and the Regression Fit Model of MINITAB 17.0 was carried out to assess the effectiveness of *Musa paradisiaca* (banana) peel extract as a green corrosion inhibitor for mild steel in acidic medium. The effect of concentration of inhibitor and reaction temperature was investigated while the efficiency of corrosion inhibition was evaluated by gasometric and thermometric methods. The system's response (inhibition efficiency) showed a stochastic distribution with respect to the independent variables, with the highest corrosion inhibition efficiency being 72.03 and 71.56%, for gasometric and thermometric methods, respectively. This peak performance occurred when the concentration was 10 mol/L and temperature 303 K. Furthermore, the ANOVA of the quadratic regression model for the corrosion inhibition showed the significant level of the model at 90.72 and 95.56% for gasometric and thermometric methods, respectively. The response surface as well as the contour plots indicated the extract from the agro-waste was most efficient at temperature below 307.5 K and at concentrations between 8 and 10 g/L. The least inhibitory effect was observed at temperatures above 318 K and at concentrations below 6 mol/L. Banana peel extract is one of those plant extracts that have shown to be promising in green corrosion inhibition.

Author details

Olusola S. Amodu[1*], Moradeyo O. Odunlami[1], Joseph T. Akintola[2], Seteno K. Ntwampe[3] and Seide M. Akoro[4]

1 Department of Chemical Engineering, Lagos State Polytechnic, Ikorodu, Lagos, Nigeria

2 Department of Chemical Engineering, University of Lagos, Akoka, Lagos, Nigeria

3 Bioresource Engineering Research Group (BioERG), Department of Biotechnology, Cape Peninsula University of Technology, Cape Town, South Africa

4 Department of Chemical Sciences Lagos State Polytechnic, Ikorodu, Lagos

*Address all correspondence to: amodu.s@mylaspotech.ng.edu

References

[1] NACE International. International Measures of Prevention, Application and Economics of Corrosion Technology (IMPACT) Study. 2016. Available from: http://inspectioneering.com/ [Accessed: May 23, 2018]

[2] Liu Z, Kleiner Y. State of the art review of inspection technologies for condition assessment of water pipes. Measurement. 2013;**46**(1):1-15

[3] Kishawy HA, Gabbar HA. Review of pipeline integrity management practices. International Journal of Pressure Vessels and Piping. 2010;**87**(7):373-380

[4] Iribarren JI, Liesa F, Alemán C, Armelin E. Corrosion rate evaluation by gravimetric and electrochemical techniques applied to the metallic reinforcing structures of a historic building. Journal of Cultural Heritage. 2017;**27**:153-163

[5] Gadiyar HS, Chintamani D, Gaonkar KB. Chemical cleaning, decontamination and corrosion. Bombay, India: Bhabha Atomic Research Centre; 1991

[6] Rebak RB, Perez TE. Effect of Carbon Dioxide and Hydrogen Sulfide on the Localized Corrosion of Carbon Steels and Corrosion Resistant Alloys. Paper Presented at the CORROSION 2017; 2017

[7] Snihirova D, Taryba M, Lamaka SV, Montemor MF. Corrosion inhibition synergies on a model Al-Cu-Mg sample studied by localized scanning electrochemical techniques. Corrosion Science. 2016;**112**:408-417

[8] Kim BH, Lim SS, Daud WRW, Gadd GM, Chang IS. The biocathode of microbial electrochemical systems and microbially-influenced corrosion. Bioresource Technology. 2015;**190**:395-401

[9] El Haleem SA, El Wanees SA, Bahgat A. Environmental factors affecting the corrosion behaviour of reinforcing steel. VI. Benzotriazole and its derivatives as corrosion inhibitors of steel. Corrosion Science. 2014;**87**:321-333

[10] Tan Y-J, Bailey S, Kinsella B. An investigation of the formation and destruction of corrosion inhibitor films using electrochemical impedance spectroscopy (EIS). Corrosion Science. 1996;**38**(9):1545-1561

[11] Taneja JN. Dissolution of iron in hydrochloric acid [Electronic Theses and Dissertations]. Canada: University of Windsor; 1967

[12] Song G, Atrens A, StJohn D. An hydrogen evolution method for the estimation of the corrosion rate of magnesium alloys. Magnesium Technology. 2001;**2001**:254-262

[13] Volk C, Dundore E, Schiermann J, Lechevallier M. Practical evaluation of iron corrosion control in a drinking water distribution system. Water Research. 2000;**34**(6): 1967-1974

[14] Verma C, Olasunkanmi LO, Ebenso EE, Quraishi MA, Obot IB. Adsorption behavior of glucosamine-based, pyrimidine-fused heterocycles as green corrosion inhibitors for mild steel: Experimental and theoretical studies. The Journal of Physical Chemistry C. 2016;**120**(21):11598-11611

[15] Zheng X, Zhang S, Gong M, Li W. Experimental and theoretical study on the corrosion inhibition of mild steel by 1-octyl-3-methylimidazolium L-prolinate in sulfuric acid solution. Industrial & Engineering Chemistry Research. 2014;**53**(42):16349-16358

[16] Verma C, Quraishi M, Kluza K, Makowska-Janusik M, Olasunkanmi LO, Ebenso EE. Corrosion inhibition of mild steel in 1M HCl by D-glucose derivatives of dihydropyrido [2, 3-d: 6, 5-d′] dipyrimidine-2, 4, 6, 8 (1H, 3H, 5H, 7H)-tetraone. Scientific Reports. 2017;7:44432

[17] Bethencourt M, Botana F, Calvino J, Marcos M, Rodriguez-Chacon M. Lanthanide compounds as environmentally-friendly corrosion inhibitors of aluminium alloys: A review. Corrosion Science. 1998;**40**(11):1803-1819

[18] Morad M. Effect of amino acids containing sulfur on the corrosion of mild steel in phosphoric acid solutions containing Cl^-, F^- and Fe^{3+} ions: Behavior under polarization conditions. Journal of Applied Electrochemistry. 2005;**35**(9):889-895

[19] Abiola OK, Oforka N, Ebenso E, Nwinuka N. Eco-friendly corrosion inhibitors: The inhibitive action of Delonix Regia extract for the corrosion of aluminium in acidic media. Anti-Corrosion Methods and Materials. 2007;**54**(4):219-224

[20] Nnaji NJ, Ujam OT, Ibisi NE, Ani JU, Onuegbu TO, Olasunkanmi LO, et al. Morpholine and piperazine based carboxamide derivatives as corrosion inhibitors of mild steel in HCl medium. Journal of Molecular Liquids. 2017;**230**:652-661

[21] Odewunmi N, Umoren S, Gasem Z. Watermelon waste products as green corrosion inhibitors for mild steel in HCl solution. Journal of Environmental Chemical Engineering. 2015;**3**(1):286-296

[22] Ituen E, James A, Akaranta O, Sun S. Eco-friendly corrosion inhibitor from *Pennisetum purpureum* biomass and synergistic intensifiers for mild steel. Chinese Journal of Chemical Engineering. 2016;**24**(10):1442-1447

[23] Ekpe U, Ibok U, Ita B, Offiong O, Ebenso E. Inhibitory action of methyl and phenyl thiosemicarbazone derivatives on the corrosion of mild steel in hydrochloric acid. Materials Chemistry and Physics. 1995;**40**(2):87-93

[24] Gerengi H, Sahin HI. *Schinopsis lorentzii* extract as a green corrosion inhibitor for low carbon steel in 1 M HCl solution. Industrial & Engineering Chemistry Research. 2011;**51**(2):780-787

[25] Muthumanickam S, Jeyaprabha B, Karthik R, Elangovan A, Prakash P. Adsorption and corrosion inhibiting behavior of *Passiflora foetida* leaf extract on mild steel corrosion. International Journal of Corrosion and Scale Inhibition. 2015;**4**(4):365-381

[26] Alibakhshi E, Ramezanzadeh M, Bahlakeh G, Ramezanzadeh B, Mahdavian M, Motamedi M. *Glycyrrhiza glabra* leaves extract as a green corrosion inhibitor for mild steel in 1 M hydrochloric acid solution: Experimental, molecular dynamics, Monte Carlo and quantum mechanics study. Journal of Molecular Liquids. 2018;**255**:185-198

[27] Amodu OS, Ntwampe SK, Ojumu TV. Emulsification of hydrocarbons by biosurfactant: Exclusive use of agrowaste. BioResources. 2014a;**9**(2):3508-3525

[28] Mohapatra D, Mishra S, Sutar N. Banana and its by-product utilisation: An overview. Journal of Scientific and Industrial Research. 2010;**69**:323-329

[29] ASTM. Standard Test Method for Kinematic Viscosity of Transparent and Opaque Liquids (the Calculation of Dynamic Viscosity). West Conshohocken, PA: ASTM International; 1997. www.astm.org

[30] Ebenso E, Eddy N, Odiongenyi A. Corrosion inhibition and adsorption properties of methocarbamol on mild

steel in acidic medium. Portugaliae Electrochimica Acta. 2009;**27**(1):13-22

[31] Okafor P, Ebenso E. Inhibitive action of *Carica papaya* extracts on the corrosion of mild steel in acidic media and their adsorption characteristics. Pigment & Resin Technology. 2007;**36**(3):134-140

[32] Maneerat N, Tangsuphoom N, Nitithamyong A. Effect of extraction condition on properties of pectin from banana peels and its function as fat replacer in salad cream. Journal of Food Science and Technology. 2017;**54**(2):386-397

[33] SPE. Oil and gas facilities: Corrosion monitoring and mitigation technologies. Society of Petroleum Engineering. 2013;**6**(2):1-6

[34] Kliškić M, Radošević J, Gudić S, Katalinić V. Aqueous extract of *Rosmarinus officinalis* L. as inhibitor of Al–Mg alloy corrosion in chloride solution. Journal of Applied Electrochemistry. 2000;**30**(7):823-830

[35] Betiku E, Akintunde AM, Ojumu TV. Banana peels as a biobase catalyst for fatty acid methyl esters production using Napoleon's plume (*Bauhinia monandra*) seed oil: A process parameters optimization study. Energy. 2016;**103**:797-806

[36] Noor EA, Al-Moubaraki AH. Corrosion behavior of mild steel in hydrochloric acid solutions. International Journal of Electrochemical Science. 2008;**3**(1):806-818

[37] Gunavathy N, Murugavel S. Corrosion inhibition studies of mild steel in acid medium using *Musa acuminata* fruit peel extract. Journal of Chemistry. 2012;**9**(1):487-495

[38] Mayanglambam RS, Sharma V, Singh G. *Musa Paradisiaca* extract as a green inhibitor for corrosion of mild steel in 0.5 M sulphuric acid solution. Portugaliae Electrochimica Acta. 2011;**2011, 29**(6):405-417

[39] Lai C, Xie B, Zou L, Zheng X, Ma X, Zhu S. Adsorption and corrosion inhibition of mild steel in hydrochloric acid solution by S-allyl-O, O′-dialkyldithiophosphates. Results in Physics. 2017;7:3434-3443

[40] Amodu OS, Ntwampe SK, Ojumu TV. Optimization of biosurfactant production by *Bacillus licheniformis* STK 01 grown exclusively on *Beta vulgaris* waste using response surface methodology. BioResources. 2014b;**9**(3):5045-5065

[41] Akalezi CO, Enenebaku CK, Oguzie EE. Inhibition of acid corrosion of mild steel by biomass extract from the Petersianthus macrocarpus plant. Journal of Materials and Environmental Science. 2013;**4**(2):217-226

[42] Raja PB, Qureshi AK, Rahim AA, Osman H, Awang K. *Neolamarckia cadamba* alkaloids as eco-friendly corrosion inhibitors for mild steel in 1 M HCl media. Corrosion Science. 2013;**69**:292-301

[43] Negm N, Kandile N, Aiad I, Mohammad M. New eco-friendly cationic surfactants: Synthesis, characterization and applicability as corrosion inhibitors for carbon steel in 1 N HCl. Colloids and Surfaces A: Physicochemical and Engineering Aspects. 2011;**391**(1-3):224-233

[44] Gunasekaran G, Chauhan L. Eco friendly inhibitor for corrosion inhibition of mild steel in phosphoric acid medium. Electrochimica Acta. 2004;**49**(25):4387-4395

[45] Verma C, Quraishi M, Ebenso EE, Bahadur I. A green and sustainable approach for mild steel acidic corrosion inhibition using leaves extract: Experimental and DFT studies. Journal of Bio-and Tribo-Corrosion. 2018;**4**(3):33

[46] Ong CC, Karim KA. Inhibitory effect of red onion skin extract on the corrosion of mild steel in acidic medium. Chemical Engineering Transactions. 2017;**56**:913-918

[47] Bao J, Zhang H, Zhao X, Deng J. Biomass polymeric microspheres containing aldehyde groups: Immobilizing and controlled-releasing amino acids as green metal corrosion inhibitor. Chemical Engineering Journal. 2018;**341**:146-156

Chapter 4

Green Corrosion Inhibitory Potentials of Cassava Plant (*Manihot esculenta* Crantz) Extract Nanoparticles (CPENPs) in Coatings for Oil and Gas Pipeline

Funsho O. Kolawole, Shola K. Kolawole, Oluwamayowa M. Olugbemi and Suleiman B. Hassan

Abstract

Internal and external corrosion affects oil and gas pipelines and were discussed in this chapter. Corrosion inhibitors are one of the methods that can be used to achieve corrosion control and prevention. The main discussion in this chapter was the use of cassava plant (*Manihot esculenta* Crantz) extract nanoparticles (CPENPs) as an additive in coatings to serve as a green corrosion inhibitor for oil and gas pipeline. Trace elements, such as O, Si, Ca, K, Fe and S, which are hetero-atoms, have been identified in CPENPs. Elements like Si and Ca would also improve the strength of coatings as well as reduce corrosion rate of coated metals. It has also been revealed that CPENPs is composed of the following compounds SiO_2, $CaCO_3$, $Ca_2(SO_4)_2H_2O$ and $CaC_2O_4(H_2O)$, which would help in improving the mechanical properties of alloys, composites and coatings. SiO_2 if added to coatings will improve the coating hardness, while the presence of $CaCO_3$ in coatings will form a precipitate that will serve as a protective film on the surface of the metal, thereby protecting the metal from corrosion. The nature of bond and organic compounds that exist in the CPENPs was also discussed.

Keywords: cassava extract, coatings, corrosion, green inhibitor, nanoparticles

1. Introduction

Corrosion is the main problem affecting pipelines in the oil and gas industry. Internal corrosion in oil and gas pipelines is primarily caused by the presence of water together with acid gases or sulphate reducing bacteria. It can be categorized into three: sweet corrosion, sour corrosion and microbiological influenced corrosion. Conversely, in external corrosion the medium in the surrounding reacts with the outer side of metal pipelines thereby causing certain damages. The soil is complex three-phased system, which makes it a conductor to the metal pipelines. The application of coating and inhibitors has help to solve the corrosion problem. Coatings may be applied alone or may be used with other common methods such

IntechOpen

as proper material selection, cathode protection (CP) and application of inhibitors to modify the corrosive environment. In time past different types of coatings have been used such as bituminous enamels, asphalt mastic, liquid epoxies and phenolics, extruded plastic coatings, fusion-bonded epoxy (FBE), tape, three-layer polyolefin, wax coatings, high performance composite coating system, low temperature application technology for powder, thermotite flow assurance coating technology and paint. Paints used as coatings have been very effective in reducing the rate of corrosion in many industries, where various metals which are prone to corrosion are put into use.

Use of some inhibitors, such as chromate, has been banned because of toxicity and the environmental hazards they create [1]. Hence there is a need to make use of environmental friendly, non-toxic extracts of naturally occurring plant materials as corrosion inhibitors also known as green corrosion inhibitors. Extracts of plant materials contain a wide variety of organic compounds. Most of them contain hetero-atoms such as phosphorous, nitrogen, sulfur and oxygen. These atoms coordinate with the corroding metal atom (their ions) through their electrons [2]. Hence protective films are formed on the metal surface, thereby preventing corrosion. Some of the green corrosion inhibitors that have been used in the past are extracts from Aloe vera, Banana Plant juice, Mango, Orange, Passion fruit, Cashew peels, Tobacco leaves, etc. [1, 3–5].

Most researches have led to the discovering of new green corrosion inhibitors from plant part, but few of them have provided practical coating application that can solve corrosion problems as it applies to the oil and gas pipelines. Most green corrosion inhibitors are extracted from food and fruit items e.g. mango, orange, cassava, cashew juice and bitter leaf, which endangers food security. It is important to identify the ingredients and nature of bonds that exist in plant part extracts which are responsible for inhibiting corrosion in steel pipelines. The use of liquid extracts from cassava tuber has been studied in the past for corrosion resistance for mild steel [6]. But there is still work to be done in the area of coating using green inhibitors. Cassava plant solid extract (bark, stem and leaf) which are dumped in the environment can be processed and used as corrosion inhibitors because of the presence of heteroatoms and organic compounds, which are responsible for inhibiting corrosion. Processing of cassava plant solid extract (bark, leaf and stem) into green corrosion inhibitor will provide a cheap, environmental friendly and efficient green corrosion inhibitor and also solve environmental challenges caused by the dumped cassava waste.

Paint which usually consists of four main parts includes: solvents, pigments, binders and additives. In dealing with anticorrosive paints, red oxide pigment helps in preventing corrosion. The additives in paints play vital roles, which include the inhibition of corrosion. Inhibitors can be added to paint and it must be non-toxic, environmentally friendly, cost effective and readily available. Extracts from plants have been used as green inhibitors in the past and have been found to be environmentally friendly, non-toxic, cheap and abundantly available. Nanoparticles can improve paint property because of the unique properties of nanoparticles. Incorporating nanosize additives in paint provide effective barrier performance, reduce the amount of holiday in paint and also enhance the integrity and durability of paint, since the fine particles dispersed in paint can fill cavities. Research has been carried out on cassava starch as green corrosion inhibitors [7], and the modification of cassava starches as potential corrosion inhibitors has been used on steel rods. Cassava starch which is sometimes a waste in most cassava processing plants, shows considerable improvement in the prevention of corrosion [8].

This current chapter seeks to discuss green corrosion inhibitory potentials of cassava plant (*Manihot esculenta Crantz*) extract nanoparticles (CPENPs) in coatings for oil and gas pipeline. The cassava extract to be considered here are the bark, leaf and stem, which is usually dumped all over the environment as waste. Cassava extract can be processed into nanoparticles and added into coatings as a green corrosion inhibitor.

2. Corrosion problems in oil and gas pipeline

Corrosion is the main problem affecting oil and gas pipelines. Understanding the electro-chemical nature of corrosion was a major breakthrough, as shown in **Figure 1**, and this made it possible for corrosion to be mitigated, if electric current sufficient to offset the inherent corrosion current of a particular environment were caused to flow in the opposite direction. The applied direct current was termed "cathodic" protection because it made the pipe the cathode in a galvanic cell [10, 11]. The required current could be supplied by connecting a "sacrificial" anode (i.e., a metal with a higher oxidation potential than iron) in an electrical circuit where soil acted as the "electrolyte." Alternatively, commercial current could be directed to the pipe via an anode bed. The application of coating and inhibitors also help to solve the corrosion problem [10, 11].

2.1 Internal corrosion in pipeline

Internal corrosion in oil and gas pipelines is primarily caused by the presence of water together with acid gases (carbon dioxide or hydrogen sulphide) or sulphate reducing bacteria [12]. It can be divided into three broad categories:

- Sweet corrosion
- Sour corrosion
- Microbiological influenced corrosion (MIC)

2.2 External corrosion in pipeline

The medium in the surrounding reacts with the outer side of metal pipelines chemically, electrochemically and physically causing certain damages. These damages are called external corrosion of pipelines. The soil is complex three-phased

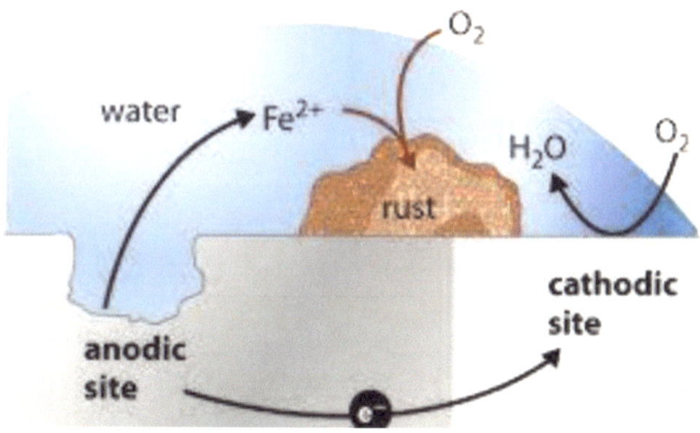

Figure 1.
Schematic of anodic site and cathodic site as they lead to corrosion [9].

system, which makes it a conductor to the metal pipelines. Plus, the oxygen concentration cell caused by the oxygen concentration difference accelerates the pipeline corrosion [13].

- Differential cell corrosion

- Galvanic corrosion

2.3 Control and prevention of corrosion in the oil and gas industry

Control and prevention of corrosion in the oil and gas industries has been on for a long time. Although many methods have been suggested to arrest corrosion, they can be classed broadly into four main categories; selection of appropriate materials, use of inhibitors, use of protective coatings and cathodic protection [13].

3. Corrosion inhibitors

An inhibitor is a substance that, when added in small concentrations to an environment, decreases the corrosion rate. In a sense, an inhibitor can be considered as a retarding catalyst. There are numerous inhibitor types and compositions and are generally classified into organic and inorganic inhibitors (**Figure 2**). Most inhibitors have been developed by empirical experimentation, and many inhibitors are proprietary in nature and thus their composition is not disclosed. The chemicals, their concentration, and the frequency of injection depend on the process medium and, normally, on the recommendations of the inhibitor manufacturer, since these chemicals, although generic in nature, are generally proprietary items [10, 13, 15]. The inhibitors used are normally chromates, phosphates, and silicates, added following the recommendations of the manufacturer. The removal of oxygen from a fluid medium improves the chances of corrosion resistance by materials in contact with the fluid. Controlling and stabilizing the pH value of the medium is another method of combating corrosion Inhibition is not completely understood because of these reasons, but it is possible to classify inhibitors according to their mechanism and composition. The corrosion rates of usefully resistant materials generally range between 1 and 200 mpy [10, 13, 15].

3.1 Inorganic corrosion inhibitors

Substances, such as arsenic and antimony ions, specially retard the hydrogen-evolution reaction. As a consequence, these substances are very effective in acid

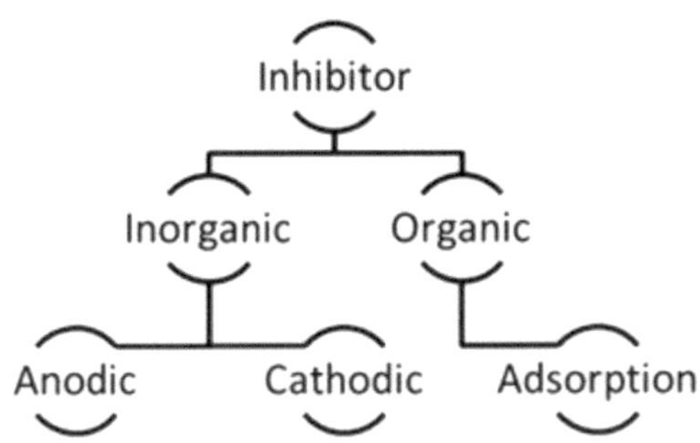

Figure 2.
Classification of inhibitors [14].

solutions but are ineffective in environments where other reduction processes such as oxygen reduction are the controlling cathodic reactions [15]. Scavengers are substances that act by removing corrosive reagents from solution. Oxidizers are substances as chromate, nitrate, and ferric salts also act as inhibitors in many systems. In general, they are primarily used to inhibit the corrosion of metals and alloys that demonstrate active-passive transitions, such as iron and its alloys and stainless steels [15]. Vapor-phase inhibitors are very similar to the organic adsorption-type inhibitors and possess a very high vapor pressure. Generally, the inorganic inhibitors have cathodic actions or anodic.

3.1.1 Anodic inhibitors

Anodic inhibitors (also called passivation inhibitors) act by a reducing anodic reaction, that is, blocks the anode reaction and supports the natural reaction of passivation metal surface, also, due to the forming a film adsorbed on the metal. In general, the inhibitors react with the corrosion product, initially formed, resulting in a cohesive and insoluble film on the metal surface [14, 16, 17]. The anodic inhibitors reacts with metallic ions Me^{n+} produced on the anode, forming generally, insoluble hydroxides which are deposited on the metal surface as insoluble film and impermeable to metallic ion, from the hydrolysis of inhibitors results in OH^- ions [16]. Some examples of anodic inorganic inhibitors are nitrates, molybdates, sodium chromates, phosphates, hydroxides and silicates [14, 15]. Potentiostatic polarization diagram showing electrochemical behavior of a metal in a solution with anodic inhibitor (a) versus without inhibitor (b) is illustrated in **Figure 3**.

3.1.2 Cathodic inhibitors

As the corrosion process begins, the cathodic corrosion inhibitors prevent the occurrence of the cathodic reaction of the metal. These inhibitors have metal ions able to produce a cathodic reaction due to alkalinity, thus producing insoluble compounds that precipitate selectively on cathodic sites. Deposit over the metal a compact and adherent film, restricting the diffusion of reducible species in these areas. Thereby, increasing the impedance of the surface and the diffusion restriction of the reducible species, in this case, the oxygen diffusion and electrons conductive in these areas. These inhibitors cause high cathodic inhibition [14, 16–18]. **Figure 4** shows an example of a polarization curve of the metal on the solution with a cathodic inhibitor. When the cathodic reaction is affected the corrosion, potential

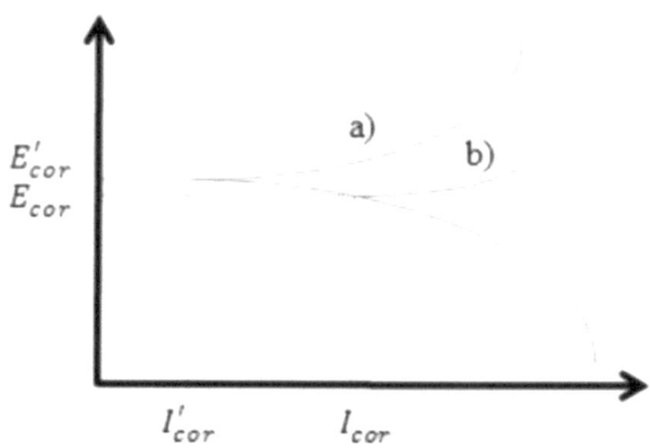

Figure 3.
Potentiostatic polarization diagram: electrochemical behavior of a metal in a solution with anodic inhibitor (a) versus without inhibitor (b) [14].

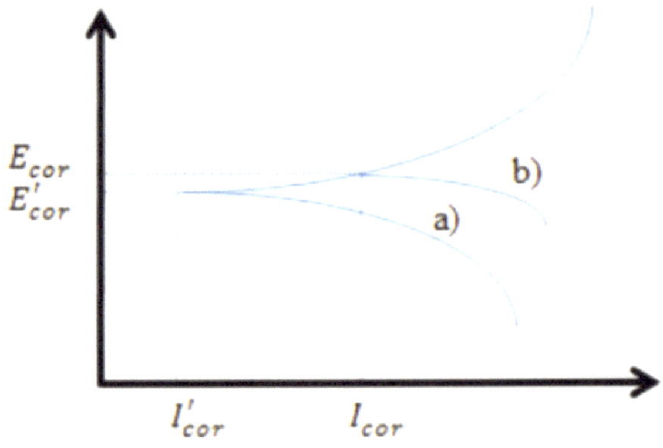

Figure 4.
Potentiostatic polarization diagram: electrochemical behavior of a metal in a cathodic inhibitor solution (a), as compared to the same solution without inhibitor (b) [14].

is shifted to more negative values [15]. When cathodic inhibitors minimize the release of hydrogen ions due to a phenomenon that can difficult the discharge of the hydrogen, called overvoltage [14].

3.2 Organic corrosion inhibitors

Environmental concerns require corrosion inhibitors to be nontoxic and environment friendly and acceptable. Green chemistry serves as a source of environmental friendly green corrosion inhibitors. Corrosion inhibitors are extensively used in corrosion protection of metals and equipment. Organic compounds with functional groups containing nitrogen, sulfur, and oxygen atoms are generally used as corrosion inhibitors. Most of these organic compounds are not only expensive but also harmful to the environment. Thus, efforts have been directed toward the development of cost effective and nontoxic corrosion inhibitors. Plant products and some other sources of organic compounds are rich sources of environmentally acceptable corrosion inhibitors. An example of such a system is the corrosion inhibition of carbon steel by caffeine in the presence and absence of zinc. Plant products are a source of environment-friendly green inhibitors such as phthalocyanines [2]. After the addition of the inhibitor, the corrosion potential remains the same, but the current decreases from I_{cor} to I'_{cor}. Is showed in **Figure 5** the mechanism of actuation of organic inhibitors, when it is adsorbed to the metal surface and forms a protector film on it.

The inhibitor efficiency could be measured by the follow equation:

$$Ef = \frac{R_i - R_o}{R_o} x\,100 \quad (1)$$

where E_f is inhibitor efficiency (percentage), R_i is corrosion rate of metal with inhibitor and R_o is corrosion rate of metal without inhibitor [14].

3.3 Corrosion inhibitor mechanism

Corrosion inhibition mechanisms operating in an acid medium differs widely from one operating in a near-neutral medium. Corrosion inhibition in acid solutions can be achieved by halides, carbon monoxide, and organic compounds containing functional group heteroatoms such as nitrogen, phosphorus, arsenic, oxygen, sulfur, and selenium, organic compounds with multiple bonds, proteins, polysaccharides, glue, bitumen, and natural plant products such as chlorophyll and anthocyanins [19]. The initial step in the corrosion inhibition of metals in acid solutions consists of adsorption of the inhibitor on the oxide-free metal surface followed by retardation of the cathodic and/or the anodic electrochemical corrosion reactions [19]. Corrosion inhibitors work by forming a protective film on the metal preventing corrosive elements contacting the metal surfaces, as illustrated in **Figure 6**.

The action mechanisms of corrosion inhibitors are;

- By adsorption, forming a film that is adsorbed onto the metal surface,
- By inducing the formation of corrosion products such as iron sulfide, which is a passivizing species,
- By changing media characteristics, producing precipitates that can be protective and eliminating or inactivating an aggressive constituent.

It is well known that organic molecules inhibit corrosion by adsorption, forming a barrier between the metal (pipeline) and the environment. Thus, the polar group of the molecule is directly attached to metal and the non-polar end is oriented in a vertical direction to the metal surface, which repels corrosive species as illustrated in **Figure 6**, furthermore establishing a barrier against chemical and electrochemical attack by fluids on the metallic surface [20, 21].

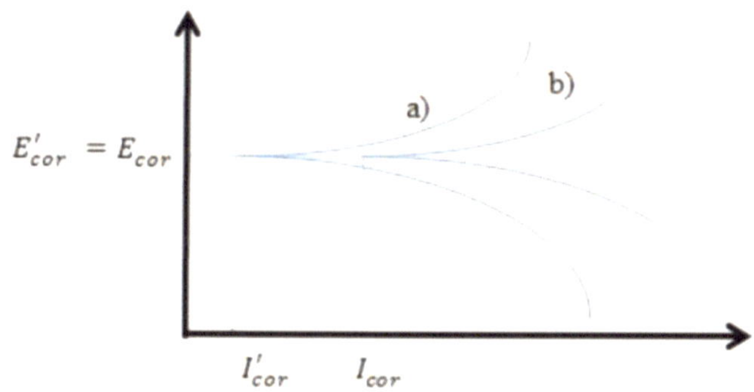

Figure 5.
Theoretical potentiostatic polarization diagram: electrochemical behavior of a metal in a solution containing a cathodic and anodic inhibitor (a) compared to the same solution without the inhibitor (b) [14].

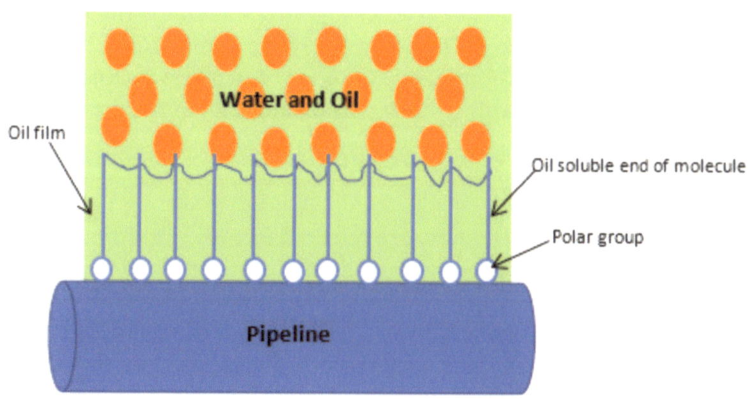

Figure 6.
Schematic of modified inhibitor film prevents water contacting a pipeline surface.

An inhibitor may be effective in one system, while in another it is not, therefore, it is convenient to consider the following factors: chemical structure of the inhibitor component, chemical composition of the corrosive medium, nature of the metal surface (pipeline), operating conditions (temperature, pressure and pH) thermal stability of the inhibitor—corrosion inhibitors have temperature limits above which lose their effectiveness because they suffer degradation of the containing components, solubility of the inhibitor in the system—the solubility of the inhibitor in the system is required to achieve optimum results in the metal surface protection; this depends on the length of the hydrocarbon chain, the addition of surfactants to enhance the dispersibility or solubility of inhibitors, and modification of the molecular structure of the inhibitor by ethoxylation to increase the polarity, and thus reach its solubility in the aqueous medium [21].

The main features of an inhibitor are:

- Ability to protect the metal surface.
- High activity to be used in small quantities (ppm).
- Low cost compounds.
- Inert characteristics to avoid altering a process.
- Easy handling and storage.
- Preferably with low toxicity.
- Non-contaminant.
- It should act as an emulsifier.
- It should act as a foaming agent [21].

4. Paints for corrosion protection

Paints are made up of a mixture of different components, although paints designed for different purposes will have different formulations, they all have some key features in common. Paints contain a pigment to give color, including white; a film former that binds the pigment particles together and binds them to the surface to be painted; a liquid that makes it easier to apply the paint and additives to make the basic paint better to store and to use. There are two main types of paint, which are gloss and emulsion [22]. **Table 1** shows a typical gloss paint formulation.

4.1 Alkyd resin binder

The alkyd resins produced this way are referred to as oil-modified alkyd resins and contribute about 70% to the conventional binders used in surface coating [23]. They determine the performance quality of surface coatings such as the rate of drying, gloss, durability of the dry film and resistance of the dry film to abrasion and chemicals. However, classification of alkyd resins is based on the oil length and oil type [24]. The vegetable oils used in oil-modified alkyd resins are usually extracted either by mechanical press or solvent extraction [25]. The natural oil in the oil-modified alkyds reacts with atmospheric oxygen leading to the formation of

Component	Mass (%)
Alkyd resin binder	30
Pigment	25
Solvent	40
Additives	5

Table 1.
A typical gloss paint formulation [22].

network of polymers cross-linked through the C=C bond. The oxidative drying of the oil brings about the formation of film that shows improved properties with drying time, hardness or water resistance [26]. The oils used in surface coatings contain linolenic and conjugated acid groups, such oils include linseed, perilla and tung oils and possess pronounced drying abilities [27]. There has been tremendous increase in the demand for alkyd resin production for use in the Nigerian surface coating industry due to the rapid growth of the economy [27].

4.2 Red oxide pigment

The use of iron oxides as natural pigments has been practiced since earliest times. The iron oxides such as magnetite, hematite, maghemite and goethite are commonly used as pigments for black, red, brown and yellow colors respectively. Predominantly natural red iron oxides are used in primers for steel constructions and cars reducing corrosion problems. Iron oxide are strong absorbers of ultraviolet radiation and mostly used in automotive paints, wood finishes, construction paints, industrial coatings, plastic, nylon, rubber and print ink [28].

4.3 Solvent

Solvents are not the only means of removing low molecular weight compounds. Heat can help evaporate saturated fatty acids, such as palmitic and stearic acids, and an improperly stored or displayed painting can become embrittled by the loss of these plasticizers. The long-term behavior of oil paints also seems to indicate that a small amount of evaporation of fatty acids occurs over time. Improper temperature on a hot table may do so as well since the volatility of the fatty acids becomes significant above 70–80°C [29].

The mechanical properties of a paint film depend upon its basic structure and the presence of small organic molecules that may act as plasticizers. The original structure of a paint film contains the ester bonds of the oil and the bonds produced by the cross-linking of the unsaturated fatty acids through autoxidation. The loss of any of these bonds results in weakening the film strength. Any loss of the ester bonds must have a significant effect on the structure of the oil paint. After 6 years, the paints made with varying degrees of hydrolyzed oil appear as coherent films, but some disintegrate when solvents, such as acetone or toluene, are applied because these solvents can remove the low molecular weight compounds that contribute to the stability of the paints [29].

4.4 Additives

Additives are small amounts of substances that modify the paint properties, additives might be driers anti-skin agents, anti-corrosive agents, antifreeze, dispersing aids, wetting agents, thickeners, biocides, low temperature drying aids, anti-foam agent, and coalescing solvent. Driers accelerate the paints drying (hardening) by catalyzing the oxidation of the binder, while plasticisers increase the paints flexibility. Fungicides, biocides and insecticides prevent growth and attack of fungi, bacteria and insects and flow control agents improve flow properties. Defoamers prevent formation of air bubbles entrapped in the coatings, emulsifiers are wetting agents increasing the colloidal stability of the paints in liquid state, while UV stabilizers provide stability of the paints under ultra-violet light, and anti-skinning agents prevent formation of a skin in the can. Adhesion promoters improve the adhesion of the coating to the substrate, and texturizers impart textures to the coatings [29].

4.5 Nanoparticles

Nanoparticles are important scientific tools that have been and are being explored in various biotechnological, pharmacological and pure technological uses. They are a link between bulk materials and atomic or molecular structures. While bulk materials have constant physical properties regardless of its size, among nanoparticles the size often dictates the physical and chemical properties. Thus, the properties of materials change as their size approaches the nanoscale and as the percentage of atoms at the surface of a material becomes significant. For bulk materials, those larger than 1 μm (or micron), the percentage of atoms at the surface is insignificant in relation to the number of atoms in the bulk of the material. Nanoparticles are unique because of their large surface area and this dominates the contributions made by the small bulk of the material [30, 31].

In typical nanomaterials, the majority of the atoms located on the surface of the particles, whereas they are located in the bulk of conventional materials. Thus, the intrinsic properties if nanomaterials are different from conventional materials since the majority of atoms are in a different environment. Nanomaterials represent almost the ultimate in increasing surface area and they are chemically very active because the number of surface molecules or atoms is very large compared with the molecules or atoms in the bulk of the materials. Substances with high surface areas have enhanced physical, chemical, mechanical, optical and magnetic properties and this can be exploited for a variety of structural and non-structural application. Nanoparticles/fillers find application in wear-resistant, erosion-resistant and corrosion resistant [31]. Coatings with nanostructure bring about a reduction in surface contact tension, minimization of moisture penetration, and reduction in surface roughness to 1 nm for better dirt repellence [31].

4.6 Synthesis of nanoparticles

Nanoparticles of various types has been synthesized; gold, silver, magnetite, zinc oxide, silicon oxide, and others [32–35], which can be synthesized using methods such as the breakdown (top-down) method and the build-up (bottom-up) method [33].

5. Plant extracts

Plants naturally synthesize chemical compounds in defense against fungi, insects and herbivorous mammals. Some of these compounds or phytochemicals such as alkaloids, terpenoids, flavonoids, polyphenols and glycosides prove beneficial to humans in unique manner for the treatment of several diseases. These compounds are identical in structure and function to conventional drugs. Extracts from parts of plants such as roots, stems, and leaves also contain such extraordinary phytochemicals that are used as pesticides, antimicrobials, drugs and herbal medicines [4, 36–38].

5.1 Plant extracts as green inhibitors

Plant extracts are excessively used as corrosion inhibitors. Plant extracts contain a variety of organic compounds such as alkaloids, flavonoids, tannins, cellulose and polycyclic compounds. The compounds with hetero atoms-N, O, S, P coordinate with (corroding) metal atom or ion consequently forming a protective layer on the metal surface, which prevents corrosion. These serve as cheaper, readily available, renewable and environmentally benign alternatives to costly and hazardous corrosion inhibitors (e.g., chromates). Plant extracts serve as anticorrosion agents to various metals such as mild steel, copper, zinc, tin, nickel, aluminum and its alloys [37, 38].

There are exhaustive numbers of plant extracts that have shown proven anticorrosion activity as corrosion inhibitors. Examples are *Swertia angustifolia*, *Acacia concinna*, *Emblica officinalis*, *Terminalia chebula*, *Terminalia belivia*, *Sapindus trifoliatus*, *Pongamia glabra*, Eucalyptus leaves, *Annona squamosa*, *Eugenia jambolana*, *Azadirachta indica*, *Acacia arabica*, *Vernonia amygdalina*, *Carica papaya*, *Rosmarinus officinalis*, *Hibiscus sabdariffa*, Opuntia extracted, *Mentha pulegium*, *Ocimum viridis*, *Datura metel*, *Ricinus communis*, *Chelidonium majus*, Papaia, *Poinciana pulcherrima*, *Cassia occidentalis* and *Datura stramonium* seeds, Papaia, *Calotropis procera* B, *Azydracta indica*, *Justicia gendarussa*, *Artemisia pallens*, *Auforpio turkiale* sap, Black pepper extract, henna extract and several others [4, 36–38].

5.2 Cassava plant

Cassava can be grown on a wide range and can yield satisfactorily even in acidic soils where most other crops fails [39], the crop has continually played very vital roles, which include income for farmers, low cost food source for both the rural and urban dwellers as well as household food security. In Nigeria, Cassava is generally believed to be cultivated by small scaled farmers with low resources. It also plays a major role in the effort to alleviate the food crisis in Africa [39]. Cassava with botanical name *Manihot esculenta*, is a woody shrub of the spurge family, Euphorbiaceae, native to South America. It is extensively cultivated as an annual crop in tropical and subtropical regions for its edible starchy tuberous root, a major source of carbohydrates [40]. World production of cassava root was estimated to be 245 million tonnes in 2012 [41]. Africa produces about 137 million tonnes, which is the largest contribution to the world production; 75 million tonnes is produced from Asia; and 33 million tonnes in Latin America and the Caribbean, specifically Jamaica. Nigeria is the world's largest producer of cassava, producing about 37.5 million tonnes annually [41]. A mature cassava root (hereafter referred to as 'root') may range in length from 15 to 100 cm and weigh 0.5–2.5 kg. Circular in cross-section, it is usually fattest at the proximal end and tapers slightly toward the distal portion. It is connected to the stem by a short woody neck and ends in a tail similar to a regular fibrous root.

5.2.1 Cassava bark (CB)

Cassava bark also known as cassava peels are always dumped in abundant as waste. Although, it has been reported that both leaf and bark contains cyanogenic glucosides, linamarin, lotaustralin, starch, amino acid, carbohydrate, proteins and tannin [41]. Cassava bark (CB) consist of two layers namely; the outer skin and inner skin [42], both layers combine together serves as agro-waste and since annual production is high and it has been reported that the bark consist of 5–10% of the cassava root [42–44], the amount of agro-waste that can be generated from cassava bark is significant. CB is used for animal feed [42], biogas [43].

5.2.2 Cassava leaf (CL)

Cassava leaves are sometimes considered as agro-waste, though it has other applications such as animal feed [42, 45], medicinal application (Aro, 2008) and pack-cyaniding of mild steel [45]. Cassava leaves are also known for their high HCN content, low energy, bulkiness and their high tannin content [42, 45]. Cassava leaves are nutritionally valuable products and cassava plant could yield 7–15 tonnes of leaves per hectare, which accounts for an additional 1 tonne of valuable protein and 2.5 tonnes of carbohydrate per hectare [42]. Up to 6% of cassava leaves can be obtained from the total production of cassava [42].

5.2.3 Cassava stem (CS)

Cassava stem is the largest waste generated from cassava plantation after harvest, up to 400 bundles can be obtained per hectare (Information and Communication Support for Agricultural Growth in Nigeria, 2015). From the estimated amount of cassava stem waste generated it shows that enough can be obtained for processing into useful application. Cassava stem can be fed to pigs, poultry, dairy cattle [45] and biochar production [46].

5.3 Synthesis of cassava plant extract nanoparticles (CPENPs)

CPENPs which comprises of cassava bark nanoparticles (CBNPs) [47], cassava leaf nanoparticles (CLNPs) [48] and cassava stem nanoparticles (CSNPs) [48] were obtained by first soaking for 24 h, after which ball milling for 60 h was carried out, to achieve a particles size below 100 nm which were estimated by SEM/Gwyddion software, XRD and TEM [47, 48]. Trace elements such as O, Si, Ca, K, Fe and S, were revealed using EDX, which are hetero-atoms and can be added to coatings to help in inhibiting corrosion on metal surfaces [47, 48]. Elements like Si and Ca would improve the strength of coatings as well as reduce corrosion rate of coated metals [47, 48]. XRD revealed compounds such as SiO_2, $CaCO_3$, $Ca_2(SO_4)_2H_2O$ and $CaC_2O_4(H_2O)$, these compounds would help in improving the mechanical properties of alloys or composites and coatings. SiO_2 if added to coatings will improve the coating hardness, while the presence of $CaCO_3$ in coatings will form a precipitate that will serve as a protective film on the surface of the metal, thereby protecting the metal from corrosion. FTIR result revealed the nature of bond that exist in the CLNPs and GC–MS result showed various organic compounds that were presence in the CLNPs [48]. These organic compounds can be classified as fats, waxes, alkaloids, proteins, phenolics, simple sugars, pectins, mucilages, gums, resins, terpenes, starches, glycosides, saponins and essential oils. All of which helps improve the properties of metallic coatings [47, 48]. This chapter discuss the synthesis and characterization of CPENPs which can be used as additives to coatings for corrosion protection, especially coatings for oil and gas applications due to the properties as discussed by Kolawole et al. [46–48]. Therefore, CPENPs should not be left to waste as they are useful for additives in coatings. These will add value to the CPENPs that is usually dumped in the environment and also reduces environmental pollution.

6. Conclusions

Utilization of cassava plant extract (bark, leaf and stem) nanoparticles as green corrosion inhibitors incorporated into paint or coatings as an additive. The cassava plant part waste utilized will reduce the amount of waste contributing to environmental nuisance. The cassava plant solid extract (bark, leaf and stem) will serve as wealth creation to the farmers and value addition to the cassava waste. The developed corrosion resistant paint will enhance corrosion resistant of API 5 L X65 steel pipeline used in the oil and gas industries. Since the world production of cassava is about 268 million tonnes annually, the cassava waste generated will be significantly high, therefore the developed corrosion resistant paint will be cheaper and efficient because of the presence of heteroatoms and organic compounds which help in inhibiting corrosion.

Acknowledgements

The authors will like to acknowledge Dr. Abdulhakeem Bello, Physics Department, University of Pretoria, South Africa. Mr. Elakhame in the Ceramic Department in FIIRO.

Conflict of interest

Authors have no conflict of interest.

Author details

Funsho O. Kolawole[1,2]*, Shola K. Kolawole[3,4], Oluwamayowa M. Olugbemi[4] and Suleiman B. Hassan[5]

1 Department of Materials and Metallurgical Engineering, Federal University Oye-Ekiti, Ekiti State, Nigeria

2 Department of Mining and Petroleum Engineering, University of Sao Paulo, Sao Paulo, Brazil

3 National Agency for Science and Engineering Infrastructure, Nigeria

4 Department of Materials Science and Engineering, African University of Science and Technology, Abuja, Nigeria

5 Department of Metallurgical and Materials Engineering, University of Lagos, Lagos, Nigeria

*Address all correspondence to: funsho.kolawole@fuoye.edu.ng

References

[1] Eddy NO, Odoemelam SA. Inhibition of corrosion of mild steel in acidic medium using ethanol extract of Aloe vera. Pigment & Resin Technology. 2009;**38**(2):111-115

[2] Sastri VS. Green corrosion inhibitors. In: Theory and Practice. Hoboken, NJ: John Wiley & Sons; 1998

[3] Bouyanzer A, Hammouti B, Majidi L. Pennyroyal oil from *Mentha pulegium* as corrosion inhibitor for steel in M HCl. Materials Letters. 2006;**60**(23):2840-2843

[4] Satapathy AK, Gunasekaran G, Sahoo SC, Amit K, Rodrigues PV. Corrosion inhibition by *Justicia gendarussa* plant extract in hydrochloric acid solution. Corrosion Science. 2009;**51**(12):2848-2856

[5] Souza FS, Spinelli A. Caffeic acid as a green corrosion inhibitor in mild steel. Corrosion Science. 2009;**5**(3):642-649

[6] Jekayinfa SO, Waheed MA, Adebiyi KA, Adebiyi FT. Effect of cassava fluid on corrosion performance of mild steel. Anti-Corrosion Methods and Materials. 2005;**52**(5):286-292

[7] Szymońska J, Targosz-Korecka M, Krok F. Characterization of starch nanoparticles. Journal of Physics: Conference Series. 2009;**146**:012027

[8] Ochoa N, Bello M, Sancristobal J, Balsamo A, Albornoz, Brito JL. Modified cassava starches as potential corrosion inhibitors for sustainable development. Materials Research. 2013;**16**(6):1209-1219

[9] Rajput RK. Material Science and Engineering. Nai – Saraki, Delhi: S. K. Kataria & Sons; 2006. pp. 854-872

[10] Holloway MD, Nwaoha C, Onyewuenyi OA. Process Plant Equipment: Operation, Control, and Reliability. 1st ed. Hoboken, New Jersey, United States: John Wiley & Sons, Inc; 2012

[11] Kiefner JF, Trench CJ. Oil pipeline characteristics and risk factors: Illustrations from the decade of construction. American Petroleum Institute's Pipeline Committee Publication. 2001:1-59

[12] Palmer-Jones R, Paisley D. Repairing internal corrosion defects in pipelines—A case study. In: 4th International Pipeline Rehabilitation and Maintenance Conference; Prague; 2000. pp. 1-25

[13] Liu X, Wang X. The research of oil and gas pipeline corrosion and protection technology. Advances in Petroleum Exploration and Development. 2014;**7**(2):102-105

[14] Dariva CG, Galio AF. Corrosion inhibitors—Principles, mechanisms and applications. Developments in Corrosion Protection. 2014:365-379

[15] Fontana MG. Corrosion Engineering. 3rd ed. Washington, DC, United States: McGraw-Hill Book Company; 1986. pp. 1-277, 300-444

[16] Corrosão VG. Vicente Gentil. Corrosao. 6th ed. Rio de Janeiro: LTC; 2003

[17] Roberge PR. Handbook of Corrosion Engineering. New York: McGraw-Hill; 2000

[18] Talbot D, Talbot, e J. Corrosion Science and Technology. Florida: CRC Press; 2000

[19] Sangeetha M, Rajendran S, Muthumegala TS, Krishnaveni A. Green corrosion inhibitors—An overview. Zastita Materijala. 2011;**52**(1)

[20] Hobbs J. Reliable corrosion inhibition in the oil and gas industry. In: Health and Safety Executive. 2014

[21] Palou RM, Olivares-Xomelt O, Likhanova NV. Environmentally friendly corrosion inhibitors. Developments in Corrosion Protection. 2014:431-465

[22] Paints: Gloss and Emulsion. Available from: http://resources.schoolscience.co.uk/ICI/14-16/paints/paintch1pg/ [Retrieved: 2016]

[23] Bajpai M, Seth S. Use of unconventional oils in surface coating blends of alkyd resins with epoxy esters. Pigment & Resin Technology. 2000;**29**:82-87

[24] Micciche F. The Combination of Ascorbic Acid Derivative/Iron Salts as Catalyst for the Oxidative Drying of Alky-Paints: A Biomimetic Approach. Enschede, The Netherlands: Technische Universiteit Eindhoven; 2005. http://alexandria.tue.nl/extra2/200512801.pdf

[25] Abulude FO, Ogunkoya MO, Ogunleye RF. Storage properties of oils of two Nigeria oil seeds *Jatropha curcas* (physic nut) and *Helianthus annuus* (sunflower). American Journal of Food Technology. 2007;**2**:207-211

[26] Akintayo CO. Synthesis and characterization of *Albizia benth* medium oil modification of alkyd resin and its chemically modified derivatives [PhD thesis]. Ibadan, Nigeria: University of Ibadan; 2004

[27] Aigbodion AI, Okiemen FE, Obazee EO, Bakare IO. Utilization of maleinized rubber seed oil and its alkyd resin as binders in water-borne coatings. Progress in Organic Coatings. 2003;**46**:28-31

[28] Palanisamy KL, Devabharathi V, Meenakshi SN. Corrosion inhibition studies of mild steel with carrier oil stabilized of iron oxide nanoparticles incorporated into a paint. International Journal of ChemTech Research. 2015;**7**(4):1661-1664

[29] Erhardt D, Tumosa CS, Mecklenburg MF. Can artists' oil paints be accelerated aged? Polymer Preprints. 2000;**41**(2):1790-1791

[30] Mandal. Synthesis of Nanoparticles. 2012. Available from: http://www.news-medical.net/health/Synthesis-of-Nanoparticles.aspx [Accessed: February 14, 2015]

[31] Mathiazhagan A, Joseph R. Nanotechnology—A new prospective in organic coating—Review. International Journal of Chemical Engineering and Applications. 2011;**2**(4):225-237

[32] Ghandoor HE, Zidan HM, Khalil MMH, Ismail MIM. Synthesis and some physical properties of magnetite (Fe_3O_4) nanoparticles. International Journal of Electrochemical Science. 2012;**7**:5734-5745

[33] Horikoshi S, Serpone N. Microwaves in Nanoparticle Synthesis. 1st ed. Hoboken, New Jersey, United States: Wiley-VCH Verlag GmbH & Co: KGaA; 2013

[34] Lue JT. Physical properties of nanomaterials. Encyclopedia of Nanoscience and Nanotechnology. 2007;**X**:1-46

[35] Vanaja A, Rao KS. Sol-gel synthesis and characterization of pure and silver doped zinc oxide nanoparticles. International Journal of Advanced Engineering and Nano Technology. 2015;**2**(8):11-14

[36] Dahmani M, Et-Touhami A, Al-Deyab SS, Hammouti B, Bouyanzer A. Corrosion inhibition of C38 steel in 1 M HCl: A comparative study of black pepper extract and its isolated piperine.

International Journal of Electrochemical Science. 2010;**5**:1060-1069

[37] Ostovari A, Hoseinieh SM, Peikari M, Shadizadeh SR, Hashemi SJ. Corrosion inhibition of mild steel in 1 M HCl solution by henna extract: A comparative study of the inhibition by henna and its constituents (lawsone, gallic acid, ‡-D-glucose and tannic acid). Corrosion Science. 2009;**51**(9):1935-1949

[38] Zucchi F, Omar IH. Plant extracts as corrosion inhibitors of mild steel in HCl solutions. Surface Technology. 1985;**24**(4):391-399

[39] Claude F, Denis F. African cassava mosaic virus: Etiology, epidemiology, and control. Plant Disease. 1990;**74**(6):404-411. DOI: 10.1094/pd-74-0404

[40] FAOSTAT: Production, Crops, Cassava, 2010 Data. Food and Agriculture Organization; 2012

[41] Aro SO. Improvement in the nutritive quality of cassava and its by-products through microbial fermentation. African Journal of Biotechnology—Review. 2008;**7**(25):4789-4797

[42] Ofoefule AU, Uzodinma EO. Biogas production from blends of cassava (*Manihot utilissima*) peels with some animal wastes. International Journal of Physical Sciences. 2009;**4**(7):398-402

[43] Sonnenberg AS, Baars JJ, Obodai M, Asagbra A. Cultivation of oyster mushrooms on cassava waste. In: Proceedings of the 8th International Conference on Mushroom Biology and Mushroom Products; 2014. pp. 286-291

[44] Adetunji AR, Isadare DA, Akinluwade KJ, Adewoye OO. Waste-to-wealth applications of cassava—A review study of industrial and agricultural applications. Advances in Research. 2015;**4**(4):212-229

[45] Noor NM, Shariff A, Abdullah N. Slow pyrolysis of cassava wastes for biochar production and characterization. Iranica Journal of Energy & Environment. 2012;**3**(Special Issue on Environmental Technology):60-65

[46] Kolawole FO, Kolawole SK, Agunsoye JO, Bello SA, Adebisi JA, Soboyejo WO, Hassan SB. Synthesis and characterization of cassava bark nanoparticles. Materials Research Society Advance. 2018:1-8

[47] Kolawole FO, Kolawole SK, Agunsoye JO, Bello SA, Adebisi JA, Okoye OC, Hassan SB. Cassava leaf nanoparticles (CLNPs) as a potential additive to anti-corrosion coatings for oil and gas pipeline. Tribology in Industry. 2017;**39**(1):63-72

[48] Kolawole FO, Agunsoye JO, Bello SA, Adebisi JA, Hassan SB. Particle size analysis and characterization of cassava (*Manihot esculenta* Crantz) stem nanoparticles (CSNPs) via a top-down approach. Annals of Science and Technology - B. 2016;**1**(1):52-55. DOI: 10.22366/ast.2016.01.008

Chapter 5

Green Corrosion Inhibitors

Lipiar K. M. O. Goni and Mohammad A. J. Mazumder

Abstract

Corrosion is an unavoidable fact in everyday life but always receive attention to control due to its technical, economical, and esthetical importance. Corrosion inhibitors are one of the most widely used and economically viable methods protecting metals and alloys against corrosion. Typical corrosion inhibitors are bio-toxic organic compounds, which have serious issue on toxicity. Considering the toxicity of the inhibitors, there is a tremendous interest in searching for an eco-friendly, and non-toxic green corrosion inhibitor. This chapter briefly discusses the importance and different methods of corrosion inhibitors with a particular emphasis given to the discussion on the different characteristic feature of the green corrosion inhibitors reported in the literature as a comparative view of organic inhibitors.

Keywords: corrosion inhibitors, mild steel, green corrosion, organic, adsorption

1. Introduction

Generally, corrosion is regarded as the loss of a metal by the influence of corrosive agents [1]. However, in a broad sense, corrosion is the devastative consequence of chemical reaction between a metal or metal alloy and its environment [2]. General corrosion or uniform corrosion is the most prevalent form of corrosion that takes place on an entirely exposed metal surface *via* the electrochemical reactions in atmospheric or aqueous media and continues uniformly to cause the greatest destruction of that metal [3]. Even though only metals come to mind when describing corrosion, non-metallic materials, such as plastics, concrete, ceramics, rubber, etc. are prone to corrosion as well when exposed to different corrosive environments [4].

The difference in the potential energies of the corroding metal and the corrosion product is the fundamental force that drives the corrosion reaction. A certain amount of energy is required to be provided to naturally occurring minerals and ores to extract metals from them. Therefore, it is natural that these metals tend to revert back to their original state from which they were obtained when they are exposed to their environments. It is noteworthy that each metal is different in terms of the amount of energy required and stored in it or that is released during its corrosion. The greater the amount of energy needed during metal extraction, the more thermodynamically unstable is the metal and the shorter is its temporary existence in metallic form. Hence, corrosion has also been defined as the reverse of extractive metallurgy [1, 5].The electrochemical dissolution of a metal is the most important mechanism involved in its corrosion and makes the basis of all uniform and localized corrosion types. However, there are some corrosion types, such as oxidation, fretting corrosion, molten salt corrosion, etc. that can be described without reference to electrochemistry [1]. The mechanism of corrosion attack in an atmospheric environment and in an aqueous environment will be always governed by some aspect of electrochemistry. Electrons will be flowing from certain areas of a metal surface to other areas through an electrolyte that is capable of conducting ions. This stems from the incredible tendency of metals to react electrochemically with water, oxygen, and other substances in the aqueous environment. In an electrochemical corrosion process, the anode is that area of the metal surface that is corroding due to the loss of electrons while the cathode is the area that consumes the electrons generated by the corrosion reaction [6].

Almost all of us are familiar with corrosion happening to metal structures, boats, steel pilings, household utensils, etc. Unfortunately, many of us are not aware of corrosion that is deteriorating the properties of underground water, oil, and gas pipelines crisscrossing our land or water pipes in the home where corrosion occurs mostly from the inside. Successful enterprises put in considerable efforts in controlling corrosion at the design stage and in the operational phase to avoid major corrosion failures, such as unscheduled shutdowns, fatalities, personal injuries, and environmental contamination in a modern business environment. However, even the best design is unable to foresee all conditions that can allow corrosion intruding into the life of a system [5]. Steel reinforced bar (rebar) can corrode in concrete without being noticed at all and can cause damage to buildings, bridges, parking structures, the collapse of electrical towers, failure of a section of highway, etc., resulting in a huge amount of repairing cost and threatening public safety [7]. This is why regular maintenance of the metallic components that are susceptible to corrosion is of paramount importance.

2. Impact of corrosion

Even though the main reasons for considering corrosion are economic and ecological, losses due to corrosion or costs of corrosion can be actually divided into three main categories as shown in **Figure 1** [8].

2.1 Material and energy

The impact of corrosion on the equipment and its surrounding deserves a huge attention when it comes to designing an industry. Corrosion is considered to be one of the most challenging issues for most of the industrialized countries. Corrosion of tanks, piping, metal components of machines, bridges, ships, etc. can incur a massive material and economic losses upon a nation. Additionally, the safety of operating equipment, such as boilers, pressure vessels, metallic containers for toxic chemicals, bridges, turbine blades and rotors, automotive steering mechanisms, and airplane components can be threatened by corrosion failure [8]. Furthermore, the

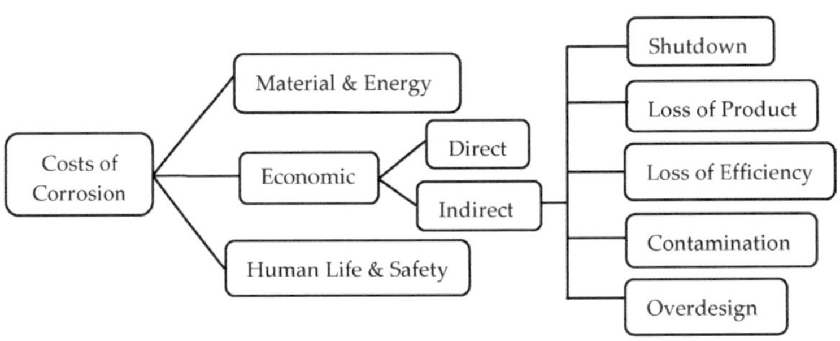

Figure 1.
Breakdown of corrosion costs.

devastative impact of corrosion goes beyond the metals and extends to energy, water, and the manufacturing phase of the metal frames [9]. It has been reported that one ton of steel turns into rust every 90 seconds, and, on the contrary, the energy needed to manufacture one ton of steel is approximately equal to the energy an average family consumes over 3 months. Approximately 50% of every ton of steel produced by the world is used to replace rusted steel [10].

2.2 Economic

Economic losses are classified into two types: (i) direct losses and (ii) indirect losses. Replacing the corroded structures and machinery of their components, for instance, mufflers, condenser tubes, pipelines, metal roofing, including necessary labor, repainting structures to prevent rusting, maintenance cost of cathodic protection system for underground pipelines, replacement cost of millions of domestic hot-water tanks and automobile mufflers, extra cost of using corrosion-resistant metals and alloys, galvanizing or nickel plating of steel, addition of corrosion inhibitors to water, and dehumidifying cost of the metal equipment storage rooms contribute to the direct losses. While it is quite difficult to assess the indirect losses, they have still been reported to add several billion dollars to the direct losses. Indirect losses include sudden shutdown of plants, loss of water, gas, or oil through a corroded pipeline, loss of efficiency in the energy conversion systems imposed by corrosion processes, contamination of water and food products in metal piping and containers, and overdesign requiring equipment to be designed many times heavier than normal operating pressure or applied stress to extend their lifetime [8]. Uhlig made the first ever systematic study on the cost of corrosion in 1949 [11]. Uhlig's report estimated the annual cost of corrosion in the United States to be US$ 5.5 billion or 2.1% of the 1949 gross national product (GNP). This study measured the total costs by summing the costs related to anti-corrosion materials and corrosion-induced maintenance and replacement handled by owners and operators (direct) as well as those related to users (indirect) [5, 12].

Corrosion cost studies using different methods, such as Uhlig method invented by Uhlig in 1949 [11], Hoar method invented by Hoar in 1971 [13], and economic input/output model devised by National Bureau of Standards (NBS) collaborating with Battelle Memorial Institute in 1978 [14] have been undertaken by several major economies, including Australia, China, Finland, Germany, India, Japan, Kuwait, the United Kingdom, and the United States [15]. A common observation of these studies was that the costs of corrosion ranged from approximately 1–5% of the GNP of each nation. The variation in the corrosion cost with respect to GNP was ascribed to the methodology used by each study and the specifics of each country [12].

A study done by National Association of Corrosion Engineers (NACE) as part of its International Measures of Prevention, Application, and Economics of Corrosion Technologies Study (IMPACT) revealed that the global cost of corrosion in 2013 was estimated to be US$ 2.5 trillion which was equivalent to 3.4% of the global GDP in that year [15]. This study utilized the World Bank economic sector and GDP data to relate the cost of corrosion studies to a global cost of corrosion. In order to address the economic sectors across the world, the global economy was divided into economic regions with similar economies (according to World Bank). These were: United States, European Region, India, Arab World (defined by the World Bank), Russia, China, Japan, Four Asian Tigers plus Macau, and Rest of the World. However, the costs estimated typically do not include environmental consequences or individual safety. It is noteworthy that receiving additional funds for corrosion studies, or updated information on these studies, more detailed and accurate global costs can be assessed.

2.3 Human life and safety

Corrosion can take a toll more than imaginable in human life and safety. This destructive phenomenon has been overlooked as the main reason of many fatal accidents for reasons of liability or simply because the evidence disappeared in the catastrophic event. The Silver Bridge collapse is one of the most dangerous and discussed corrosion accidents [16]. On December 15, 1967, this bridge connecting Point Pleasant, West Virginia and Kanauga, Ohio suddenly collapsed into the Ohio River to claim 46 lives. Stress corrosion cracking (SCC) and corrosion fatigue were determined to be responsible for this disaster. The Bhopal accident that took place in Bhopal, India on the night of the December 2–3, 1984 is one of the worst industrial accidents in terms of the lives lost and injuries. An unfortunate seepage of water (500 liters) caused by the corrosion of pipelines, valves, and other safety equipment into a methylisocyanate (MIC) storage tank at Union Carbide India Limited caused the release of MIC and other toxic reaction products into the surrounding areas that killed 3000 people and injured an estimated 500,000 people [17]. Swimming Pool Roof Collapse is another infamous corrosion accident that took place in Uster, Switzerland in 1985. The roof of this swimming pool was supported by stainless steel rods that failed due to SCC and killed 12 people [18].

3. Techniques for corrosion measurement

3.1 Weight loss measurement

Weight loss analysis is known to the simplest, most reliable, and long-established method of assessing corrosion losses in plant and equipment. A sample of metal or alloy under experiment is weighed and then immersed into a corrosive solution, and later removed from the corrosive medium after a predetermined time interval. The metal specimen is then weighed again after cleaning all corrosion products. The corrosion rate, surface coverage (θ), and corrosion inhibition efficiency (η%) can be calculated using (Eqs. (1)–(3)):

$$\text{Corrosion rate}\left(\frac{mm}{year}\right) = 8.76 \times 10^3 \frac{m_i - m_f}{S\rho t} \tag{1}$$

$$\theta = \frac{CR_o - CR}{CR_o} \tag{2}$$

$$\eta\% = \frac{CR_o - CR}{CR_o} \times 100\% \tag{3}$$

where m_i is the weight of the metal sample in grams before immersion, m_f is the weight of the metal sample in grams after immersion, S is the total area of metal in cm^2 that has been exposed to corrosive solution, ρ is the density of metal sample in g/cm^2, t is the time in hours during which the sample was immersed, $CR_○$ and CR represent the corrosion rates (in mmpy) without and with the inhibitor, respectively [19, 20].

3.2 Polarization measurements

Since corrosion is a phenomenon that involves electrochemistry, electrochemical-based corrosion measuring experiments provide valuable information about the rate of corrosion and mechanism of corrosion protection [20]. Polarization methods are based on changing the current or potential on a sample under investigation and recording the corresponding potential or current change. This can be facilitated with the help of either a direct current (DC) or an alternating current (AC) source [5]. Some important and widely used techniques have been discussed briefly below.

3.2.1 Tafel extrapolation

Tafel curve is a current-potential plot that shows the anodic and cathodic reactions in the electrochemical cell. In this method, the potential of the working electrode (metal sample) varied over a range at a specific rate and the resulting response in current is recorded. The anodic and cathodic reactions that are taking place simultaneously produce a total current that is represented by a curved line. The linear portions of logarithmic Tafel plot are extrapolated to produce an intersection that generates a point that signifies an approximation of the corrosion current (i_{corr}) and the corrosion potential (E_{corr}). This i_{corr} facilitates the calculation of the corrosion rate based on (Eq. (4)) and the corrosion inhibition efficiency (η%) based on (Eq. (5)):

$$\text{Corrosion rate} = \frac{i_{corr} \times K \times EW}{\rho \times A} \tag{4}$$

$$\eta\% = \frac{i^o_{corr} - i_{corr}}{i^o_{corr}} \times 100\% \tag{5}$$

where K is a constant that represents the units for the corrosion rate, EW is the equivalent weight in gram/equivalent, ρ is density in g/cm^3, A is the area in cm^2 exposed to corrosive solution, i^o_{corr} and i_{corr} are the corrosion currents in amperes without and with the inhibitor, respectively.

3.2.2 Linear polarization resistance (LPR)

Because the linear polarization resistance (LPR) method is non-destructive [21], it has quick application and can be used in the field test through portable instrumentation [22], it is the most popular of the electrochemical techniques [23]. The principal of LPR is based upon introducing a small perturbative DC electrical signal to disturb the corrosion equilibrium on the surface of metal specimen. The response of the equilibrium to this perturbation is measured with respect to a reference half-cell [24]. The polarization resistance (R_p) of a material is known to be the $\Delta E/\Delta i$ slope of a potential-current density curve at the free-corroding potential. The polarization resistance can be related to the corrosion current (i_{corr}) using the Stern-Geary approximation in which the anodic (b_a) and cathodic (b_c) Tafel slopes can be experimentally obtained from real polarization plots.

3.2.3 Electrochemical impedance spectroscopy (EIS)

Electrochemical impedance spectroscopy (EIS) is a very powerful electrochemical technique that has wide applications in the evaluation of coatings in corrosion research. This technique provides valuable information about the corrosion protection imparted by an inhibitor. In this technique, an AC voltage (in the case of potentiostatic EIS) or current (in the case of galvanostatic EIS) is applied to the system under investigation to receive response in the form of AC current (voltage) or voltage (current) as a function of the frequency. This technique can be performed in a 2- or 3-electrodes system with the help of a potentiostat-galvanostat and a frequency response analyzer (FRA) [25]. Usually, an AC voltage having small perturbations ranging from 5 to 10 mV is applied in the system over a range of frequencies typically starting from 100 kHz to 10 mHz. Based on the shape of the Nyquist plot produced by the experiment, the electrochemical cell containing the metal sample, adsorbed inhibitors, and the electrolyte medium is represented by an equivalent circuit that includes information about the solution resistance R_s, charge transfer resistance R_{ct}, and the double layer capacitance C_{dl}. A large R_{ct} value and decreasing C_{dl} values with increasing inhibitor concentrations indicate better corrosion protection [26].

4. Corrosion inhibitors

A corrosion inhibitor is known as a chemical constituent that can diminish or prevent and control corrosion when added in small amount to the metal environment. Corrosion inhibitors are considered as the first line of defense against oil and chemical industry corrosion [27]. Corrosion inhibitors are sought after giving metals temporary protection during transportation and storage as well as localized protection to prevent corrosion that may have resulted from accumulation of small amounts of an aggressive phase. An effective corrosion inhibitor should be cost-effective, compatible with the corrosive medium, and produce desired effect when present in small concentrations [28]. Corrosion inhibitors act by (i) forming a film that is adsorbed on the metal surface, (ii) producing corrosion products, for example, iron sulfide (FeS) that acts as a passivator, and (iii) yielding precipitates that can eliminate or inactivate an aggressive constituent [29].

Depending on which electrochemical reactions are being blocked, these film-forming or interface inhibitors can be classified into anodic, cathodic, or mixed-type [28, 30]. Anodic inhibitors, alternately known as passivation inhibitors, suppress the rate of anodic reactions by producing sparingly soluble deposits, such as hydroxides, oxides, or salts in close to neutral conditions. On the other hand, cathodic inhibitors function by reducing the rate of cathodic or reduction reactions by producing a protective layer on cathodic areas against hydrogen in acidic conditions and oxygen in alkaline conditions. Mixed inhibitors influence both the anodic and cathodic reaction sites by forming an adsorptive film on the metal surface. About 80% of organic inhibitors fall into this category. Based on the chemical nature of the inhibitors, they can be divided into organic and inorganic [31]. Organic and inorganic inhibitors, based on their compositions and mechanism of actions, can be further classified into neutralizing, scavenging, barrier or film-forming, and other miscellaneous inhibitors [32].

4.1 Organic inhibitors

Organic inhibitors act through forming a film on the surface of the metals and they can act as anodic, cathodic, or mixed inhibitors. The formation of this protective film happens with the help of strong interactions, such as π-orbital adsorption, chemisorption, and electrostatic adsorption that prevent the corrosive species from attacking the metal surface [33]. This adsorption is usually one molecular layer thick and does not penetrate into the bulk of the metal itself [34]. Physicochemical properties, such as functional groups, steric factors, aromaticity, π-orbital character of donating electrons, electron density at the donor atoms, and the electronic structure of the molecules govern the adsorption process [35, 36]. The corrosion inhibition efficiency of an organic inhibitor relies on its adsorption ability and mechanical, structural, and chemical characteristics of the adsorption layers formed under a specific environment [37]. An efficient organic inhibitor will usually contain polar functional groups with S, O, or N atoms in the molecule and a hydrophobic moiety that will repel the aqueous corrosive species away from the metal surface. However, the polar head is considered to be responsible for establishing the adsorption layer [38]. Some chemical families of organic inhibitors are pyridines, fatty amides, imidazolines, and 1,3-azoles [39].

4.2 Inorganic inhibitors

Inorganic inhibitors are those inhibitors in which the active substance is an inorganic compound. The addition of electropositive metal salts to a corrosive medium is one of the simplest ways to improve the passivity of a metal. However, the protective metal ion must have a redox potential more positive than the one to be protected and potentially more positive than that required for discharging protons so that the protective metal ion can be discharged on the surface of the metal in need of protection. Cathodic depolarization by overvoltage reduction and subsequent formation of an adherent deposit take place through the deposition of the protective metal on the surface of the metal susceptible to corrosion. Some of the metals that serve this purpose are palladium (Pd), platinum (Pt), iridium (Ir), rhodium (Rh), mercury (Hg), and rhenium (Re). Many inorganic anions, such as chromates (CrO_4^{2-}), molybdate (MoO_3^-), silicates (SiO_4^{4-}), phosphate ($H_2PO_3^-$), and nitrate (NO_2^-) as well provide passivation protection to the metal surfaces through their incorporation into the oxide layer [39].

Environmental friendliness, cost, availability, and toxicity are some factors that should play an extremely important role when it comes to choosing an inhibitor for a particular condition [40]. The toxicity, biodegradability, and bioaccumulation of conventional corrosion inhibitors discharged into the environment are matter of huge concern. Even though the environmental implications of commercial corrosion inhibitors are not fully understood, it is not unknown that their chemical components have hazardous impact [41]. Inorganic inhibitors, for example, arsenates, phosphates, chromates, and dichromates not only have shown promising inhibition efficiency but also have been proved intolerant as well due to the threat they pose to our social health in the long run [42]. Likewise, the ecological and health risks associated with the organic inhibitors have pushed us towards finding or using non-toxic or green corrosion inhibitors that would impart maximum protection to the metallic structures but have least impact on mankind and nature [36].

4.3 Green corrosion inhibitors

Corrosion inhibitors are extensively used for the protection of metals and equipment and they are required to be acceptable, non-toxic, and eco-friendly due to environmental concerns. The cost and harmful effect associated with the commercial organic and inorganic inhibitors have raised considerable awareness in the field of corrosion mitigation. Thus, corrosion scientists and engineers are more inclined towards the implication of green corrosion inhibitors that are inexpensive, readily available, environmentally friendly and ecologically acceptable, and renewable. Several classes of such inhibitors have been discussed briefly below.

4.3.1 Plant extracts

Umoren et al. investigated the inhibition efficiency (IE) of gum arabic (GA) in absence and presence of halide ions on mild steel in 0.1 M H_2SO_4 at different temperatures. In weight loss analysis, 0.05 M KCl, 0.05 M KBr, 0.05 M KI, and GA (0.5 g/l) alone imparted IE of 27.7, 32.2, 53.6, and 37.9%, respectively, at a maximum temperature of 60°C. At the same temperature and under similar technique, GA mixed with all of these halide solutions individually showed increased efficiency of 38.7, 47.1, and 59.1%, respectively. The ion-pair interactions between the organic cations and the halide anions have contributed to increased surface coverage that eventually led to better synergistic protection [43]. The IE of GA on AA1060 type aluminum sheets with 98.5% purity was examined at 40°C. It was found that GA (0.5 g/l) showed IE of 74.2 and 75.9% measured by hydrogen evolution and thermometric methods, respectively [44]. Buchweishaija and Mhinzi [45] investigated the IE of gum exudates from *Acacia seyal* var. *seyal* and Acacia gum from *seyal* var. *seyal* on mild steel in chlorinated drinking water using potentiodynamic polarization (PP) and electrochemical impedance spectroscopy (EIS) techniques. Gum exudates showed maximum IE of 98.5% at a concentration of 1000 ppm at 30°C. On the other hand, Acacia gum showed an IE of 96.8% at an elevated temperature of 80°C at a concentration of 600 ppm.

The anticorrosive effect of a composite coating containing chitosan (CS; green matrix), oleic acid (OA), and graphene oxide (GO; nanofiller) on mild steel in 3.5 wt.% NaCl solution has been studied by Fayyad et al. [46]. The IE of the nanocomposite coating was measured by PP and EIS techniques. It was observed that oleic acid-modified chitosan/graphene oxide (CS/GO-OA) film showed corrosion resistance 100 times better than pure chitosan (CS) coating. Additionally, oxygen transmission rate (OTR) measured for the CS/GO-OA was found to decrease by 35 folds in comparison to the pure chitosan film. This decreased OTR for the CS/GO-OA coating demonstrates that an effective barrier between the metal surface and the corrosive electrolyte species was developed. Alaneme et al. [47] experimented the IE and adsorption characteristics of elephant grass (*Pennisetum purpureum*) extract on mild steel in 1 M HCl solution. At room temperature (RT), the inhibitor showed efficiency greater than 95% and increasing with increasing concentration of the extract but decreasing with increasing temperature. The presence of hydroxyl (O-H) and unsaturated (C=C) groups that have inhibitory properties were confirmed in the extract by FT-IR investigation. The scanning electronic micrographs showed significant pitting on the metal substrate that was immersed in 1 M HCl solution without the inhibitor and pitting was rarely present in the solution that contained inhibitor. The lower rate of iron dissolution in the corrosive solution that contained inhibitor was further confirmed by a higher Fe peak by energy dispersive spectroscopy (EDS) spectrum.

The inhibitory effect of hydroxyethyl cellulose (HEC) on 1018 c-steel corrosion in 3.5% NaCl solution was studied by El-Haddad using PP, EIS, and electrochemical frequency modulation (EFM) techniques [48]. The PP study revealed that HEC acted as a mixed inhibitor and the adsorption study showed that it followed Langmuir adsorption isotherm. The fact that oxygen atoms donate unshared pair of electrons to the vacant *d*-orbital of iron was established by the optimized geometry of HEC obtained by $DMol^3$ quantum chemical calculations that showed that oxygen atoms of HEC have Muliken atomic charges with higher electron densities. The IEs measured by PP, EIS, and EFM techniques were 96.7, 95.5, and 94.8%, respectively, for the inhibitor concentration of 0.5 mM at 25°C. Mobin and Rizvi [49] explored the anticorrosion behavior of xanthan gum (XG) as an eco-friendly corrosion inhibitor for mild steel in 1 M HCl at 30, 40, 50, and 60°C, respectively. At a concentration of 1000 ppm at 30°C, XG showed the maximum IE of 74.2%. The addition of very small amounts of surfactants cetylpyridinium chloride (CPC), sodium dodecyl sulfate (SDS), and Triton X-100 (TX) improved the IE. Quantum chemical calculations found the energy differences between the highest occupied molecular orbital (HOMO) and the lowest unoccupied molecular orbital (LUMO) to be 0.05 and 0.02 eV in XG alone and XG plus SDS, respectively. The smaller energy gap between HOMO and LUMO indicated better inhibition efficiency for XG plus SDS system. The formation of complex between XG and Fe^{2+} was further confirmed by UV-Visible spectroscopic measurements. Scanning electronic microscopy (SEM) study revealed an improved surface morphology of inhibited mild steel compared to uninhibited mild steel. The plant extracts form a major class of green corrosion inhibitors. Some recently reported plant extracts as green corrosion inhibitors and their IEs have been summarized in **Table 1**.

4.3.2 Amino acids

Amino acids are molecules that contain at least one carboxyl (-COOH) group and one amino ($-NH_2$) group bonded to the same carbon atom (α- or 2-carbon). Amino acids are considered as green corrosion inhibitors because they are non-toxic, biodegradable, inexpensive, soluble in aqueous media, and easy to produce at high purity. The presence of heteroatoms, such as N, O, and S and conjugated π-electrons system have made amino acids a significant class of green corrosion inhibitors thanks to their environmental aspect [60, 61]. El-Sayed investigated the anti-corrosive effect of some amino acids, such as glycine, valine, leucine, cysteine, methionine, histidine, threonine, phenylalanine, lysine, proline, aspartic acid, arginine, and glutamic acid on carbon steel in stagnant naturally aerated chloride solutions using PP and EIS techniques. All of the amino acids acted as mixed-type inhibitor while cysteine, phenylalanine, arginine, and histidine showed remarkably high corrosion inhibition efficiency at a concentration of 10 mM/dm^3. The presence

Inhibitor (concentration)	Metal/alloy	Test condition	Maximum efficiency (η%)	Test technique	References
Saraca asoca (100 mg/L)	Mild steel	0.5 M H_2SO_4, 25°C	95.5	Tafel	[50]
Sida cordifolia (500 mg/L)	Mild steel	0.5 M H_2SO_4, 25°C	99.0	Tafel	[51]
Myristica fragrans (500 mg/L)	Mild steel	0.5 M H_2SO_4, 25°C	87.8	EIS	[52]
Ginkgo (200 mg/L)	X70 steel	1 M HCl, 45°C	92.5	EIS	[53]
Eriobotrya japonica (100% v/v)	Mild steel	0.5 M H_2SO_4, 25°C	96.2	Weight loss	[54]
Turbinaria ornata (25 g/L)	Mild steel	1 M HCl, 25°C	94.5	EIS	[55]
Pongamia pinnata (100 ppm)	Mild steel	1 N H_2SO_4, 30°C	94.6	Weight loss	[56]
Prosopis juliflora (300 ppm)	Low-carbon steel	1 M HCl, 25°C	91.5	EIS	[57]
Rollinia occidentalis (1.0 g/L)	Carbon steel	1.0 M HCl, 25°C	85.7	Tafel	[58]
Xanthium strumarium (10 mL/L)	Low-carbon steel	1 M HCl, 60°C	94.8	Weight loss	[59]

Table 1.
Some recent plant extracts as green corrosion inhibitors of different metals and alloys.

of functional groups such as OH, SH, or phenyl in the backbone of the amino acid molecules help them undergo better adsorption [62]. Mobin et al. studied the inhibitory behavior of mercapto group containing amino acid L-cysteine (CYS) on mild steel in aerated and unstirred 1 M HCl solution using weight loss (WL), PP, and EIS techniques. The maximum IE of 85.6% at 30°C was achieved with an inhibitor concentration 500 ppm. The authors investigated the effect of surfactants CPC, SDS, and TX by adding them to CYS and found that the surfactants increased the IE. CYS separately and in combination with surfactants acted as mixed-type inhibitor and obeyed Langmuir's adsorption isotherm [63].

In their attempt to solve the "bronze disease", Wang et al. studied the corrosion behavior of bronze covered with CuCl patina in the presence of CYS using EIS and X-ray photoelectron spectroscopy (XPS) techniques. EIS results revealed that CYS can both inhibit the bronze substrate and stabilize the CuCl patina effectively. The IE reached the highest value of 95.3% at a CYS concentration of 5 mmol/L. The XPS investigation showed that the chemisorption of CYS on CuCl surface happened through sulfur atom in thiol and nitrogen atom in amino group [64]. Zeino et al. investigated the mechanistic study of polyaspartic acid (PASP) on mild steel in 3% NaCl solution. PASP alone showed a moderate IE of 61% at 2.0 g/L and zinc ion added PASP showed a superb IE of 97% at a reduced PASP concentration of 0.5 g/L. The authors studied the surface morphology of mild steel utilizing SEM and atomic force microscopy (AFM) techniques. Quantum calculation and Monte Carlo simulation helped them achieve molecular level insights into the complex adsorption mechanism [65]. Ituen et al. investigated the IE of N-acetyl cysteine (NAC)-based formulation on J55, mild steel, and X80 steel at different temperatures (30–90°C). NAC-based formulations showed IEs up to 91% at 90°C [66].

4.3.3 Drugs

The use of drugs as green corrosion inhibitors has been inspired by the fact that they are non-toxic, cheap, eco-friendly, and green enough to compete with other green corrosion inhibitors because most of these drugs can be synthesized from natural products [67]. The presence of heteroatoms, benzene ring, and heterocycles, such as thiophenes, pyridine, isoxazoles, etc. have made drug molecules as a promising source of green corrosion inhibitors. Recently, Ali investigated the inhibitory behavior of Candesartan drug on carbons steel (CS) in 1 M HCl acidic medium using WL, PP, EIS, and EFM techniques. The surface morphology of the inhibited CS was investigated by EDX, AMF, and SEM techniques. The inhibitor showed an IE of 79.8% at a concentration of 300 ppm [68]. Matad et al. investigated the inhibitive properties of an anti-inflammatory drug ketosulfone as green corrosion inhibitor of mild steel in 1 M HCl acidic medium using chemical and electrochemical methods. Ketosulfone imparted a maximum IE of 96.6% at 30°C for a concentration of 200 ppm. The inhibitor was found to be of mixed-type and follow Langmuir adsorption isotherm determined by polarization measurements and thermodynamic calculations, respectively [69]. Singh et al. investigated the corrosion behavior of mild steel in 1 M HCl acidic medium in presence of an expired atorvastatin drug using WL, PP, and EIS techniques [70]. The expired drug showed an amazing IE of 99.1% at a concentration of 150 ppm. The inhibitor acted as mixed-type inhibitor with predominant cathodic behavior. Dahiya et al. studied the anti-corrosive behavior of an expired drug ethambutol on mild steel in 0.5 M HCl using WL, PP, EIS, SEM, and molecular dynamics (MD) techniques. The inhibitor showed an IE more than 95% at a concentration of 100 ppm and it acted as a mixed-type inhibitor and followed Langmuir adsorption isotherm [71].

4.3.4 Rare earth (RE) metal compounds

Chromates have applications in deoxidizers, anodizing, conversion coatings, chromate-inhibited primers, wash primers, and repair processes. Environmental protection legislation was realized to prevent the use of unacceptable materials such as chromium salts. Chromium (Cr^{6+}) is highly toxic and carcinogenic [41, 72]. The health problems associated with the use of chromates and the cost associated with their safe use and disposal have led to efforts finding good alternatives such as rare earth metal compounds. After the first paper was published by Hinton et al. [73] in 1984 on the use of cerium chloride salts as corrosion inhibitor, a lot of research papers have been published by the researcher working in this area. In 1992, Hinton et al. [74] published a review paper highlighting the use of some RE compounds as green corrosion inhibitors of a wide range of metals. In this chapter, the authors pointed that these RE salts work by producing an oxide film at the cathodic sites of the metal substrate which prevent the supply of oxygen or electrons to the reduction reaction thus reducing the corrosion rate. A comprehensive review of the recent developments in corrosion inhibitors based on RE metal compounds can be gained from the two works published by Forsyth et al. [75, 76] and the book edited by Forsyth and Hinton [77].

4.3.5 Others

Surfactant corrosion inhibitors are also known as green corrosion inhibitors because they are highly efficient, cheap, and less/non-toxic. Surfactant molecules consist of a polar hydrophilic group or "head" and a non-polar hydrophobic group or "tail". In aqueous solutions, the adsorption of surfactant occurs through either chemisorption or physisorption. Critical micelle concentration (CMC) is the most important parameter when it comes to studying the corrosion inhibition by surfactants [78]. Recently, ionic liquids have gained widespread popularity as green corrosion inhibitors. By definition, ionic liquids are referred to as materials consisting of ions having melting point below 100°C. Ionic liquids due to their some fascinating properties, such as high polarity, lower melting point, low toxicity, lower vapor pressure, very high thermal, and chemical stability have applications in some other fields of chemical and chemical engineering researches as well [79].

5. Recent patents on green corrosion inhibitors

Corrosion inhibitors are an important and cost-effective way of dealing with corrosion. A huge effort to produce green inhibitors and evaluate them successfully is underway. Some recently patented green corrosion inhibitors that could indicate the importance of this class of inhibitors are included in **Table 2**.

Inhibitor system	Reference
Main chain type polybenzoxazine (MCTPB)-Chitosan (CHI) blend	[80]
Polyaspartic acid-hydroxyphosphonoacetic acid-phosphinocarboxylic acid composition	[81]
Polyaspartic acid-soluble Sn (II) or Sn (IV) compound blend	[82]
At least (one fatty acid-one alkanolamine-one alkylamine-one organic sulfonic acid composition)	[83]
A corn stillage product	[84]

Table 2.
Some recent patents on green corrosion inhibitors.

6. Conclusion

Corrosion is a destructive phenomenon that could strike badly at the heart of an economy as it did on many occasions. The direct and indirect global costs associated with corrosion are no longer ignorable. Corrosion is not a necessary curse because it might be almost impossible to stop it but preventable. The use of inhibitors to mitigate corrosion is an economically viable option given the maintenance costs associated with the metals and equipment is already very high. However, recent trends in the field of corrosion inhibition are more inclined towards finding not only an effective inhibitor but also an eco-friendly one. This is because an effective inhibitor may serve the purpose of mitigating corrosion well but could leave hazardous impact on the health and ecology due to the harmful aspect associated with its chemical formula. Green corrosion inhibitors in the form of plant extracts, drugs, amino acids, rare earth metal compounds, ionic liquids, surfactants, etc. are not only eco-friendly but also provide excellent corrosion prevention. The replacement of the existing harmful organic and inorganic inhibitors that are used for different industrial applications by the green ones is not merely a matter of concern anymore but an agenda that has been in implementation and will continue to be realized.

Acknowledgements

The authors gratefully acknowledge the financial support and research facilities provided by Deanship of Scientific Research, King Fahd University of Petroleum and Minerals through internal project # IN131047.

Conflict of interest

The authors declare that there is no conflict of interest regarding the publication of this article.

Author details

Lipiar K. M. O. Goni and Mohammad A. J. Mazumder*
Chemistry Department, King Fahd University of Petroleum & Minerals, Dhahran, Saudi Arabia

*Address all correspondence to: jafar@kfupm.edu.sa

References

[1] Schofield MJ. Corrosion. In: Snow DA, editor. Plant Engineer's Reference Book. 2nd ed. Oxford: Butterworth Heinemann; 2002. pp. 961-986 10.1016/B978-075064452-5/50088-2

[2] Jones DA. Principles and Prevention of Corrosion. 2nd ed. New Jersey: Prentice Hall; 1996. 572 p. ISBN: 0-13-359993-0

[3] Makhlouf ASH. Intelligent stannate-based coatings of self-healing functionality for magnesium alloys. In: Tiwari A, Rawlins J, Hihara LH, editors. Intelligent Coatings for Corrosion Control. 1st ed. Oxford: Butterworth Heinemann; 2015. pp. 537-555. DOI: 10.1016/B978-0-12-411467-8.00015-5

[4] Zarras P, Stenger-Smith JD. Smart inorganic and organic pretreatment coatings for the inhibition of corrosion on metals/alloys. In: Tiwari A, Rawlins J, Hihara LH, editors. Intelligent Coatings for Corrosion Control. 1st ed. Oxford: Butterworth Heinemann; 2015. pp. 59-91. DOI: 10.1016/B978-0-12-411467-8.00003-9

[5] Roberge PR. Corrosion Engineering: Principles and Practice. 1st ed. New York: McGraw Hill; 2008. 734p. DOI: 10.1036/0071482431

[6] Corrosion Electrochemistry. 2018. Available from: https://corrosion-doctors.org/Electrochemistry-of-Corrosion/Introduction.htm [Accessed: June 12, 2018]

[7] Davis J. Corrosion. In: Understanding the Basics. 1st ed. Ohio: ASM International; 2000. 517p. ISBN: 0-87170-641-5

[8] Revie RW, Uhlig HH. Corrosion and Corrosion Control: An Introduction to Corrosion Science and Engineering. 4th ed. New Jersey: Wiley; 2008. 512p. ISBN: 978-0-471-73279-2

[9] Raja PB, Ismail M, Ghoreishiamiri S, Mirza J, Ismail MC, Kakooei S, et al. Reviews on corrosion inhibitors: A short view. Chemical Engineering Communications. 2016;**203**:1145-1156. DOI: 10.1080/00986445.2016.1172485

[10] Javaherdashti R. How corrosion affects industry and life. Anti-Corrosion Methods and Materials. 2000;**47**:30-34. DOI: 10.1108/00035590010310003

[11] Uhlig HH. The cost of corrosion to the United States. Corrosion. 1950;**6**:29-33. DOI: 10.5006/0010-9312-6.1.29

[12] Hou B, Li X, Ma X, Du C, Zhang D, Zheng M, et al. The cost of corrosion in China. Materials Degradation. 2017;**1**(4):1-10. DOI: 10.1038/s41529-017-0005-2

[13] Hoar TP. Review lecture: Corrosion of metals: Its cost and control. Proceedings of the Royal Society of London. Series A: Mathematical and Physical Sciences. 1976;**348**:1-18. http://www.jstor.org/stable/79113

[14] Bennett LH, Kruger J, Parker RL, Passaglia E, Reimann C, Ruff AW, et al. Economic Effects of Metallic Corrosion in the United States: A Report to the Congress by the National Bureau of Standards. National Bureau of Standards. Washington D. C, USA: U. S. Government Printing Office; 1978. Publication No. 511-1. 65p. [Google Books]

[15] Koch G, Varney J, Thompson N, Moghissi O, Gould M, Payer J. International Measures of Prevention, Application, and Economics of Corrosion Technologies Study. Houston: NACE International; 2016. 216p. [IMPACT Report]

[16] Lichtenstein AG. The silver bridge collapse recounted. Journal of Performance of Constructed Facilities.

1994;7:249-261. DOI: 10.1061/(ASCE)0887-3828(1993)7:4(249)

[17] Bowonder B. The Bhopal accident. Technological Forecasting and Social Change. 1987;**32**(2):169-182. DOI: 10.1016/0040-1625(87)90038-2

[18] Stress Corrosion Cracking Failure [Internet]. 2018. Available from: https://corrosion-doctors.org/Forms-SCC/swimming.htm [Accessed: July 15, 2018]

[19] El-Haddad MN. Chitosan as a green inhibitor for copper corrosion in acidic medium. International Journal of Biological Macromolecules. 2013;**55**:142-149. DOI: 10.1016/j.ijbiomac.2012.12.044

[20] Tiu BDB, Advincula RC. Polymeric corrosion inhibitors for the oil and gas industry: Design principles and mechanism. Reactive and Functional Polymers. 2015;**95**:25-45. DOI: 10.1016/j.reactfunctpolym.2015.08.006

[21] Zou Y, Wang J, Zheng YY. Electrochemical techniques for determining corrosion rate of rusted steel in seawater. Corrosion Science. 2011;**53**:208-216. DOI: 10.1016/j.corsci.2010.09.011

[22] Law DW, Millard SG, Bungey JH. Linear polarization resistance measurements using a potentiostatically controlled guard ring. NDT and E International. 2000;**33**:15-21. DOI: 10.1016/S0963-8695(99)00015-8

[23] Andrade C, Alonso C. Corrosion rate monitoring in the laboratory and on-site. Constrion and Building Materials. 1996;**10**:315-328. DOI: 10.1016/0950-0618(95)00044-5

[24] Sadowski L. New non-destructive method for linear polarisation resistance corrosion rate measurement. Archves of Civil and Mechanical Engineering. 2010;**10**:109-116. DOI: 10.1016/S1644-9665(12)60053-3

[25] Orazem ME, Tribollet B. Electrochemical Impedance Spectroscopy. 1st ed. New Jersey: Wiley-Blackwell; 2008. 525p. DOI: 10.1002/9780470381588

[26] EIS-Electrochemical Impedance Techniques [Internet]. 2018. Available from: https://www.gamry.com/application-notes/EIS/potentiostatic-eis-tutorial/ [Accessed: February 25, 2018]

[27] Taghavikish M, Dutta NK, Choudhury NR. Emerging corrosion inhibitors for interfacial coating. Coatings. 2017;**7**(12):217-245. DOI: 10.3390/coatings7120217

[28] Papavinasam S. Corrosion Inhibitors. In: Revie RW, editor. Uhlig's Corrosion Handbook. 3rd ed. New York: Wiley; 2000. pp. 1021-1033. DOI: 10.1002/9780470872864.ch71

[29] Umoren SA, Ekanem UF. Inhibition of mild steel corrosion in H_2SO_4 using exudate gum from pachylobus edulis and synergistic potassium halide additives. Chemical Engineering Communications. 2010;**197**:1339-1356. DOI: 10.1080/00986441003626086

[30] Dariva CG, Galio AF. Corrosion Inhibitors—Principles, Mechanisms and Applications. In: Aliofkhazraei M, editor. Developments in Corrosion Protection. 1st ed. Rijeka: InTech; 2014. pp. 365-381. DOI: 10.5772/57255

[31] Rahuma MN, BK M. Corrosion in oil and gas industry: A perspective on corrosion inhibitors. Materials Science and Engineering. 2014;**3**(3):110-110. DOI: 10.4172/2169-0022.1000e110

[32] Garverick L. Corrosion in the Petrochemical Industry. 2nd ed. Ohio: ASM International; 1994. 509p. [Google Books]

[33] Kelland MA. Production chemicals for the oil and gas industry.

Chromatographia. 2010;**72**(1-2):199-199. DOI: 10.1365/s10337-010-1557-2

[34] Arthur DE, Jonathan A, Ameh PO, Anya C. A review on the assessment of polymeric materials used as corrosion inhibitor of metals and alloys. International Journal of Industrial Chemistry. 2013;**4**(2):1-9. DOI: 10.1186/2228-5547-4-2

[35] Granese SL, Rosales BM, Oviedo C, Zerbino JO. The inhibition action of heterocyclic nitrogen organic compounds on Fe and steel in HCl media. Corrosion Science. 1992;**33**(9):1439-1453. DOI: 10.1016/0010-938X(92)90182-3

[36] Sabirneeza AAF, Geethanjali R, Subhashini S. Polymeric corrosion inhibitors for iron and its alloys: A review. Chemical Engineering Communications. 2015;**202**:232-244. DOI: 10.1080/00986445.2014.934448

[37] Eddy NO, Ebenso EE, Ibok UJ. Adsorption, synergistic inhibitive effect and quantum chemical studies of ampicillin (AMP) and halides for the corrosion of mild steel in H_2SO_4. Journal of Applied Electrochemistry. 2010;**40**(2):445-456. DOI: 10.1007/s10800-009-0015-z

[38] Yaro AS, Khadom AA, Wael RK. Apricot juice as green corrosion inhibitor of mild steel in phosphoric acid. Alexandria Engineering Journal. 2013;**52**:129-135. DOI: 10.1016/j.aej.2012.11.001

[39] Palou RM, Olivares-Xomelt O, Likhanova NV. Environmentally friendly corrosion inhibitors. In: Sastri VS, editor. Green Corrosion Inhibitors: Theory and Practice. 1st ed. New Jersey: Wiley; 2011. pp. 257-303. DOI: 10.1002/9781118015438.ch7

[40] Popoola L, Grema A, Latinwo G, Gutti B, Balogun A. Corrosion problems during oil and gas production and its mitigation. International Journal of Industrial Chemistry. 2013;**4**(35):1-15. DOI: 10.1186/2228-5547-4-35

[41] Sastri VS, editor. Green Corrosion Inhibitors: Theory and Practice. 1st ed. New Jersey: Wiley; 2011. 310p. DOI: 10.1002/9781118015438

[42] Obot IB, Obi-Egbedi NO, Umoren SA, Ebenso EE. Synergistic and antagonistic effects of anions and ipomoea invulcrata as green corrosion inhibitor for aluminium dissolution in acidic medium. International Journal of Electrochemical Science. 2010;**5**(7):994-1107. ISSN: 1452-38981. www.electrochemsci.org

[43] Umoren SA, Ogbobe O, Igwe IO, Ebenso EE. Inhibition of mild steel corrosion in acidic medium using synthetic and naturally occurring polymers and synergistic halide additives. Corrosion Science. 2008;**50**:1998-2006. DOI: 10.1016/j.corsci.2008.04.015

[44] Umoren SA, Ebenso EE, Okafor PC, Ogbobe O, Oguzie EE. Gum arabic as a potential corrosion inhibitor for aluminium in alkaline medium and its adsorption characteristics. Anti-Corrosion Methods and Materials. 2006;**53**(5):277-282. DOI: 10.1108/00035590610692554

[45] Buchweishaija J, Mhinzi GS. Natural products as a source of environmentally friendly corrosion inhibitors: The case of gum exudate from *Acacia seyal* var. seyal. Portugaliae Electrochimica Acta. 2008;**26**(3):257-265. DOI: 10.4152/pea.2008032257

[46] Fayyad EM, Sadasivuni KK, Ponnamma D, Al-Maadeed MAA. Oleic acid-grafted chitosan/graphene oxide composite coating for corrosion protection of carbon steel. Carbohydrate Polymers. 2016;**151**:871-878. DOI: 10.1016/j.carbpol.2016.06.001

[47] Alaneme KK, Olusegun SJ, Alo AW. Corrosion inhibitory properties of elephant grass (*Pennisetum purpureum*) extract: Effect on mild steel corrosion in 1 M HCl solution. Alexandria Engineering Journal. 2016;**55**:1069-1076. DOI: 10.1016/j.aej.2016.03.012

[48] El-Haddad MN. Hydroxyethylcellulose used as an eco-friendly inhibitor for 1018 c-steel corrosion in 3.5% NaCl solution. Carbohydrate Polymers. 2014;**112**:595-602. DOI: 10.1016/j.carbpol.2014.06.032

[49] Mobin M, Rizvi M. Inhibitory effect of xanthan gum and synergistic surfactant additives for mild steel corrosion in 1 M HCl. Carbohydrate Polymers. 2016;**136**:384-393. DOI: 10.1016/j.carbpol.2015.09.027

[50] Saxena A, Prasad D, Haldhar R, Singh G, Kumar A. Use of *Saraca asoca* extract as green corrosion inhibitor for mild steel in 0.5 M H_2SO_4. Journal of Molecular Liquids. 2018;**258**:89-97. DOI: 10.1016/j.molliq.2018.02.104

[51] Saxena A, Prasad D, Haldhar R, Singh G, Kumar A. Use of *Sida cordifolia* extract as green corrosion inhibitor for mild steel in 0.5 M H_2SO_4. Journal of Environmental Chemical Engineering. 2018;**6**:694-700. DOI: 10.1016/j.jece.2017.12.064

[52] Haldhar R, Prasad D, Saxena A. *Myristica fragrans* extract as an eco-friendly corrosion inhibitor for mild steel in 0.5 M H_2SO_4 solution. Journal of Environmental Chemical Engineering. 2018;**6**:2290-2301. DOI: 10.1016/j.jece.2018.03.023

[53] Qiang Y, Zhang S, Tan B, Chen S. Evaluation of *Ginkgo* leaf extract as an eco-friendly corrosion inhibitor of X70 steel in HCl solution. Corrosion Science. 2018;**133**:6-16. DOI: 10.1016/j.corsci.2018.01.008

[54] Zheng X, Gong M, Li Q, Guo L. Corrosion inhibition of mild steel in sulfuric acid solution by loquat (*Eriobotrya japonica* Lindl.) leaves extract. Scientific Reports. 2018;**8**:1-15. DOI: 10.1038/s41598-018-27257-9

[55] Krishnan M, Subramanian H, Dahms HU, Sivanandham V, Seeni P, Gopalan S, et al. Biogenic corrosion inhibitor on mild steel protection in concentrated HCl medium. Scientific Reports. 2018;**8**:1-16. DOI: 10.1038/s41598-018-20718-1

[56] Bhuvaneswari TK, Vasantha VS, Jeyaprabha C. *Pongamia pinnata* as a green corrosion inhibitor for mild steel in 1N sulfuric acid medium. Silicon. September 2018;**10**(5):1793-1807. DOI: doi.org/10.1007/s12633-017-9673-3

[57] Fouda AS, El-Awady GY, El Behairy WT. *Prosopis juliflora* plant extract as potential corrosion inhibitor for low-carbon steel in 1 M HCl solution. Journal of Bio- and Tribo-Corrosion. 2018;**4**:1-12. DOI: 10.1007/s40735-017-0124-x

[58] Alvarez PE, Fiori-Bimbi MV, Neske A, Brandán SA, Gervasi CA. Rollinia occidentalis extract as green corrosion inhibitor for carbon steel in HCl solution. Journal of Industrial and Engineering Chemistry. 2018;**58**:92-99. DOI: 10.1016/j.jiec.2017.09.012

[59] Khadom AA, Abd AN, Ahmed NA. *Xanthium strumarium* leaves extracts as a friendly corrosion inhibitor of low carbon steel in hydrochloric acid: Kinetics and mathematical studies. South African Journal of Chemical Engineering. 2018;**25**:13-21. DOI: 10.1016/j.sajce.2017.11.002

[60] El Ibrahimi B, Jmiai A, Bazzi L, El Issami S. Amino acids and their derivatives as corrosion inhibitors for metals and alloys. Arabian Journal of

Chemistry. 23 September 2018. DOI: 10.1016/j.arabjc.2017.07.013. (In Press)

[61] Zhang DQ, Xie B, Gao LX, Cai QR, Joo HG, Lee KY. Intramolecular synergistic effect of glutamic acid, cysteine and glycine against copper corrosion in hydrochloric acid solution. Thin Solid Films. 2011;**520**:356-361. DOI: 10.1016/j.tsf.2011.07.009

[62] El-Sayed NH. Corrosion inhibition of carbon steel in chloride solutions by some amino acids. European Journal of Chemistry. 2016;7(1):14-18. http://www.eurjchem.com/index.php/eurjchem/article/view/1331

[63] Mobin M, Zehra S, Parveen M. L-Cysteine as corrosion inhibitor for mild steel in 1 M HCl and synergistic effect of anionic, cationic and non-ionic surfactants. Journal of Molecular Liquids. 2016;**216**:598-607. DOI: 10.1016/j.molliq.2016.01.087

[64] Wang T, Wang J, Wu Y. The inhibition effect and mechanism of L-cysteine on the corrosion of bronze covered with a CuCl patina. Corrosion Science. 2015;**97**:89-99. DOI: 10.1016/j.corsci.2015.04.018

[65] Zeino A, Abdulazeez I, Khaled M, Jawich MW, Obot IB. Mechanistic study of polyaspartic acid (PASP) as eco-friendly corrosion inhibitor on mild steel in 3% NaCl aerated solution. Journal of Molecular Liquids. 2018;**250**:50-62. DOI: 10.1016/j.molliq.2017.11.160

[66] Ituen EB, Akaranta O, Umoren SA. N-acetyl cysteine based corrosion inhibitor formulations for steel protection in 15% HCl solution. Journal of Molecular Liquids. 2017;**246**:112-118. DOI: 10.1016/j.molliq.2017.09.040

[67] Fayomi OSI, Anawe PAL, Daniyan A. The impact of drugs as corrosion inhibitors on aluminum alloy in coastal-acidified medium. In: Aliofkhazraei M, editor. Corrosion Inhibitors, Principles and Recent Applications. 1st ed. Rijeka: InTech Open; 2018. pp. 79-94. DOI: 10.5772/intechopen.72942

[68] Ali AH. Electrochemical study of candesartan drug as corrosion inhibitor for carbon steel in acid medium. Journal of Advanced Electrochemistry. 2018;**4**(1):152-157. DOI: 10.30799/jaec.050.18040101

[69] Matad PB, Mokshanatha PB, Hebbar N, Venkatesha VT, Tandon HC. Ketosulfone drug as a green corrosion inhibitor for mild steel in acidic medium. Industrial & Engineering Chemistry Research. 2014;**53**:8436-8444. DOI: 10.1021/ie500232g

[70] Singh P, Chauhan DS, Srivastava K, Srivastava V, Quraishi MA. Expired atorvastatin drug as corrosion inhibitor for mild steel in hydrochloric acid solution. International Journal of Industrial Chemistry. 2017;**8**:363-372. DOI: 10.1007/s40090-017-0120-5

[71] Dahiya S, Saini N, Dahiya N, Lgaz H, Salghi R, Jodeh S. Corrosion inhibition activity of an expired antibacterial drug in acidic media amid elucidate DFT and MD simulations. Portugaliae Electrochimica Acta. 2018;**36**(3):213-213. DOI: 10.4152/pea.201803213

[72] Damborenea JD, Conde A, Arenas MA. Corrosion inhibition with rare earth metal compounds in aqueous solutions. In: Forsyth M, Hinton B, editors. Rare Earth-Based Corrosion Inhibitors. 1st ed. Cambridge: Woodhead Publishing; 2014. pp. 84-116. DOI: 10.1533/9780857093585.84

[73] Hinton BRW, Arnott DR, Ryan NE. Inhibition of aluminum alloy corrosion by cerous cations. Metals

Forum. 1984;**7**(4):211-217. [Google Scholar]

[74] Hinton BRW. Corrosion inhibition with rare earth metal salts. Journal of Alloys and Compounds. 1992;**180**:15-25. DOI: 10.1016/0925-8388(92)90359-H

[75] Forsyth M, Seter M, Tan MY, Hinton B. Recent developments in corrosion inhibitors based on rare earth metal compounds. Corrosion Engineering, Science and Technology. 2014;**49**:130-135. DOI: 10.1179/1743278214Y.0000000148

[76] Forsyth M, Seter M, Hinton B, Deacon G, Junk P. New "Green" corrosion inhibitors based on rare earth compounds. Australian Journal of Chemistry. 2011;**64**:812-819. DOI: 10.1071/CH11092

[77] Forsyth M, Hinton B, editors. Rare Earth-Based Corrosion Inhibitors. 1st ed. Cambridge: Woodhead Publishing; 2014. 338p. [Google Books]

[78] Badawi AM, Hegazy MA, El-Sawy AA, Ahmed HM, Kamel WM. Novel quaternary ammonium hydroxide cationic surfactants as corrosion inhibitors for carbon steel and as biocides for sulfate reducing bacteria (SRB). Materials Chemistry and Physics. 2010;**124**:458-465. DOI: 10.1016/j.matchemphys.2010.06.066

[79] Verma C, Ebenso EE, Quraishi MA. Ionic liquids as green and sustainable corrosion inhibitors for metals and alloys: An overview. Journal of Molecular Liquids. 2017;**233**:403-414. DOI: 10.1016/j.molliq.2017.02.111

[80] Hafiz TB, Almathami AA, Agrawal G, Hoegerl M. Green high-efficiencey corrosion inhibitor; 2018. United States Patent Application No. [US 20180057947]

[81] Drewniak M, Steimel LH. Composition and method for inhibiting corrosion and scale; 2017. United States Patent Application No. [US 20170306506]

[82] Erickson DL, Johnson RA, LaBrosse MR, Young PR. Corrosion inhibiting methods; 2016. United States Patent Application No. [US9290850B2]

[83] Meyer GR, Monk KA. Corrosion inhibitors for oil and gas applications; 2013. United States Patent Application No. [US9382467B2]

[84] Kharshan M, Fursan A, Gillette K, Kean R. Bio-based corrosion inhibitor; 2013. United States Patent Application No. [US8409340B1]

Chapter 6

Formation of Anticorrosive Structures and Thin Films on Metal Surfaces by Applying EDM

Pavel Topala, Alexandr Ojegov and Vitalie Besliu

Abstract

The methods of the formation of anticorrosive strata and thin films on metal surfaces by applying electric discharge machining (EDM) are presented in this chapter. The increase of anticorrosion resistance of metal surfaces by the formation of palladium depositions, carbon films and oxide and hydroxide films has been demonstrated. Corrosion and electrochemical behaviour of titanium with palladium powder coatings were investigated in sulfuric acid solutions at temperatures of 80 and 100°C. In order to increase the diffusion of palladium in the base material and to increase the layer homogeneity, after coating, the samples were annealed in vacuum at 1150°C for 1 hour. The proposed method allows to reduce the corrosion speed of titanium at least by two orders (from 18.7 to 0.3 g m^{-2} h^{-1}); the corrosion potential is changed towards positive values (from −0.56 mV to +0.3 V). The research on surface electrical resistance and resistance to corrosion of oxide and hydroxide films formed on steel C45 surfaces by applying EDM have shown that the surface electrical resistance of the samples increased by 10^7 times, the potential of corrosion increased from −0.44 mV to +0.4 V and the resistance to corrosion has increased by about two times in 1% NaCl water solution and by about 10 times in 30% H_2SO_4 water solution. The less pronounced increase of anticorrosion properties has carbon films formed on the same steel C45 surface; instead they increase superficial microhardness, the functional durability and processing productivity of the active piece surfaces.

Keywords: electric discharge machining, corrosion, thin films, carbon, microhardness

1. Introduction

Corrosion is a damaging process for the absolute majority of parts that work in active environments, from the chemical point of view, and becomes even more pronounced when they operate in energy fields of different types: high temperatures, light, electrical action, etc. It consists in the destruction of metallic materials under the chemical or electrochemical action of the environment or the substances they come into contact with.

With the exception of noble metals, all other metals are unstable in contact with the atmospheric air and aggressive environments. The way this instability manifests itself, and the degree to which it occurs, depends on the nature of the metal and the environment in which it is placed.

Two types of corrosion can be highlighted considering the deployment mechanism: chemical corrosion [1] which refers to the processes of destruction of metals and alloys produced in dry gases as well as in liquids without electrical conductivity and in most organic substances and electrochemical corrosion [1–8] which causes processes of metal and alloy degradation in electrolyte solutions in the presence of humidity, accompanied by the flow of electric current through the metal.

In all cases, when talking about the functionality of the machines and apparatuses or the parts from which they are assembled and their durability, the process of component degradation of different constructions is a damaging one. Throughout the practical encountered corrosion problems, it is important to know the real speed the process is taking place. If the corrosion process is possible, but has a very low deployment speed, the material is considered corrosion resistant.

Taking into account that iron and its alloys are used in the contemporary construction of machinery and apparatus as basic materials, and, on the other hand, a material for the present and future is considered to be titanium and its alloys, in the following, we will focus on the research of behaviour of these materials in different working environments.

In the case of iron, its corrosion takes place in stages by its oxidation in the atmosphere with the formation of iron oxides (rust).

In the course of the first stage of iron oxidation, FeO is formed, ferrous oxide, which is stable only in the absence of oxygen. When atmospheric oxygen appears, iron oxide is converted into iron hydroxide ($Fe_2O_3 \cdot H_2O$) or FeO(OH), of which two phases are known:

- Phase 1 that corresponds to a large excess of oxygen
- Phase 2 characterized by an insufficient amount of oxygen, which is why oxidation evolves slowly

Depending on their colour, there are three types of rust:

1. $Fe(OH)_2$ white rust, formed after the reaction:

$$Fe + 2H_2O \rightarrow Fe(OH)_2 + 2H_2$$

   This type of rust quickly passes through oxidation to brown rust, which is why it is rarely seen.

2. Brown rust occurs as a result of the following reaction

$$4Fe(OH)_2 + O_2 \rightarrow 4FeO(OH) + 2H_2O$$

3. The black rust is made of ferrous and ferric oxide, being also called magnetite because of its magnetic properties, and it is considered to be the most stable form of iron oxide. It forms a protective layer on the surface of the metal, with a homogeneous and adherent structure.

The reaction is as follows:

$$2FeO(OH) + Fe(OH)_2 \rightarrow Fe_3O_4 + 2H_2O$$

Iron and its alloys are reacted with electrochemically active acids and salts of metals with the destruction of the workpiece surface in working environments:

$$Fe + H_2SO_4 \rightarrow FeSO_4 + H_2$$
$$Fe + 2HCl \rightarrow FeCl_2 + H_2$$
$$Fe + CuSO_4 \rightarrow FeSO_4 + Cu$$
$$Fe + 2NaCl \rightarrow FeCl_2 + 2Na$$

Titanium and its alloys are more resistant to corrosion, since, under normal conditions in the atmosphere, the TiO_2 oxide film is formed, which is stable in different environments, other than those alkaline, for instance:

$$TiO_2 + Na_2CO_3\ (K_2CO_3) \rightarrow Na_2TiO_3\ (K_2TiO_3) + CO_2$$

Under certain conditions (high temperature, high concentration), titanium also reacts with some acids:

$$Ti + H_2SO_4 \rightarrow TiSO_4 + H_2$$
$$2Ti + 3H_2SO_4 \rightarrow Ti_2(SO_4)_3 + 3H_2$$
$$Ti + 8HNO_3 \rightarrow Ti(NO_3)_4 + 4NO_2 + 4H_2O$$
$$2Ti + 6HCl \rightarrow 2TiCl_3 + 3H_2$$
$$2Ti + 6HBr \rightarrow 2TiBr_3 + 3H_2$$
$$2Ti + 6HI \rightarrow 2TiI_3 + 3H_2, \text{ etc.}$$

They destroy the base metal and cause further damage of the workpieces in those environments.

The methods and techniques of anticorrosive protection are various and numerous, and they can be grouped into the following categories [9]:

- Methods of preventing corrosion.
- The use of metals and alloys resistant to corrosion.
- Methods of influencing the corrosive medium [10, 11].
- Methods of covering metal surfaces with protective layers [12–14].
- Passivation of surfaces by depositing anticorrosive strata, etc. The essence of preventing corrosion consists in [9].
- Correct choice of materials used in building devices, industrial equipment from the point of view of resistance to corrosion.
- Avoiding the contact of two metals, one of which is more electronegative than the other, e.g., aluminium and copper alloys or alloyed steel, bronze and steel, etc.
- Avoiding the contact of cold-hardened metals with annealed or cast metals, for the reason that due to the difference of electrochemical potential between them, the last one corrodes in the presence of an electrolyte.

- A more careful processing of the metal surface, since hollows and scratches favour and accelerate corrosion.

The group of metals and alloys resistant to corrosion contain noble metals and their alloys, but their use becomes difficult because of the high cost. Self-protective metals and alloys will be used instead. These are metals and alloys which, after an initial corrosion, are covered with an isolating pellicle or film as a result of the passivation phenomenon (e.g. the passivation of Ag in HCI by forming the AgCl film of Fe in HNO_3 concentrated through the formation of the $Fe(NO_3)_3$ pellicle, etc.).

In most cases, metals are alloyed with an adequate component. Sometimes the relatively low concentration of the alloyed component considerably reduces the corrosion speed (e.g., the introduction of such elements as Cu, Cr or Ni of 0.2–0.3% in steels, etc.).

There are many cases when it is possible to influence the corrosive medium to lower the speed of corrosion. There are many possibilities.

The following can be mentioned [9]:

- Modification of pH, which means bringing it to a convenient value for the metal that is to be protected. This consists in the elimination from the medium of corrosion, by the use of physical, chemical and mechanic methods (e.g., neutralization of waste water with chemical substances).

- Removal of gases (O_2 and CO_2), which increase the speed of corrosion in corrosive media, particularly that of the water.

- The use of inhibitory elements (organic or inorganic substances) [10, 11], which, being introduced in small quantities in the corrosive medium, lower or completely annihilate its corrosive capacity.

- Cathode protection (electro-defence) means the use of galvanic methods of metal protection with the help of auxiliary metal anodes, which take over the corrosive capacity of the medium, thus protecting the material used for the workpiece.

The protection realized by applying anticorrosive layers [9, 12–14] means to cover the metal with a thin layer or self-protective material. The self-protective layer should meet the following requirements: it should be compact and adherent, it should be elastic and plastic enough, and its thickness should be uniform in addition to being as small as possible. The protective layer may be metallic or nonmetallic. Metallic deposits may be realized through different methods: galvanic, thermal, through plating, laser radiation, bombarding with ions, alloying by means of electric discharge machining (EDM), etc. The formed surface layers, except those obtained through plating, usually present a certain porosity, which determines the appearance of local portions of corrosion with all the consequences. Nonmetallic protective layers may be organic or inorganic, realized with the help of varnish, paints, enamels or plastic thin sheets.

The formation of inorganic thin films or pellicles (oxide or graphite) on the workpiece surface is a progressive method of anticorrosive protection of superficial layers. The oxide or the graphite, being much more passive than the proper metal, in the interaction with the work media lowers the corrosive potential of the whole piece. The key problem that should be solved is the one that deals with the formation of structurally stable deposits with a high mechanic and electric resistance, having a uniform continuity and thickness on the whole surface.

The choice of the method to be used for the production of the protective layer depends on [9]:

- The conditions and working surroundings of the piece
- The shape and the dimensions of the protective item
- The quality of the supportive material or of the protected item
- The functional technological parameters of the device
- The way the item to be protected is placed in the device
- The technologies of application and executive possibilities for anticorrosive protection

2. Methodology of experimental investigations

2.1 Experimental setup

The experimental setup for the formation of anticorrosive coatings on the metal surfaces of the parts by applying pulsed electric discharge machining (PEDM) comprises the following electric blocks: the power pulse generator, the priming block (intended for initiating the electrical discharge) and the control block, the role of which is to synchronize the power and the priming discharge pulses.

As a result of the specialty bibliographic analysis [12, 13, 15–21], it was concluded that, for the formation of these layers, RC-type generators, with parallel priming, can be successfully used.

The deposition of layers of metallic powders with the use of electric fields is produced as a result of interaction of the powder particles with the plasma jet and the transport of the liquid and vapour phase on the surface of the piece-cathode [12, 13]. The powder particles, having the dimensions of 20–200 μm, are inserted in the 0.3–1.5 mm gap at a capacitor voltage of U_c = (80...400) V.

2.2 Methodology of anticorrosive coatings formation

Research on the interactions of the plasma channel of PEDM with the surfaces of the electrodes have shown that for the phenomenon of electroerosion, two types of effects are characterised as follows: type I, the appearance of the "cold" electrode spots on the electrode surfaces, which give rise to the asperities and impurities on their surface [12, 13] and cause both the cleaning of the surfaces of impurities and their thermal interaction, producing structural changes in the superficial layers of small thicknesses (just about the size of micro- and nanometers), and type II, on the electrode surfaces, in which, after the "cold" electrode spots, the "warm" ones cause the essential melting of the electrode, accompanied by the vaporization phenomena and the dropping of the electrode material [13]. If type II interaction of the plasma channel with electrode surfaces has found a rather wide application in dimensional processing [22–27] and the deposition layer formation (both of the compact materials [21] and the powders [12, 13]), then type I actions are described in the literature only as scientific findings, and for this reason, it is necessary to elucidate the conditions and the effects of surface thermal treatment and to reveal whether this type of interaction is purely thermal or is a thermochemical one.

Analysing the results obtained by the author of the paper [13], it was established that, in order to obtain a type I interaction with the plasma channel (in the absence of surface melting) or type II interaction (when the liquid phase is attested) on the surfaces of the workpieces, it is necessary that the energy density emitted on the processed surface be lower or higher than the specific heat of melting of the material, and the latter can be appreciated by the relation [13, 16, 28]:

$$\frac{4W}{\pi d_c^2 \cdot S} < Q < \frac{4W}{\pi d_c^2 \cdot S} \qquad (1)$$
$$Q = q\rho,$$

where q and ρ are the specific heat of melting and the density of the workpiece material, respectively, W is the energy emitted in the gap, d_c is the diameter of the plasma channel and S is the gap value.

As can be seen from the relation (Eq. 1), at the time of the energetic processing regime, the size of the gap and the thermo-physical properties of the workpiece material are known; the diameter of the plasma channel can be determined, which coincides with the size of its trace on the processed surface. If a coefficient of overlapping of the traces (of type I or type II) k = 0.5...1.0 and the frequency of the electrical impulses f are known, the productivity of the technological process can be determined by the relation:

$$\eta = \frac{k\pi d_c^2 f}{4} \qquad (2)$$

In the paper [13], it has been demonstrated that when PEDM is applied for superficial processing, the erosion processes take place, accompanied by explosive melting and vaporization of the electrode material for the vast majority of studied metals and alloys, for the current pulse duration contained in the limits 10^{-4}–10^{-9} s. It results that, in order to obtain the expected effects, it is necessary to provide a relatively low-duration pulses of discharge.

Effects on electrode surfaces are a function depending on the way of the workpiece connection in the discharge circuit (as anode or as cathode) [13, 16, 21]. These desiderata have been studied in [13], and it has been established that for short-term discharge pulses, they are “cathodic” and for the long-term ones, they are “anodic”; thus, in the case of superficial thermal treatments and deposition of materials, the workpiece will be included in the discharge circuit as cathode. In the case of thermal treatment of steel surfaces, their hardness increases by 2–3 times and for titanium surfaces by 2–5 times, for the formed layers thicknesses from a few micrometres to several dozens of micrometres. The depth of these layers reaches a maximum at three passes of technological cycle for steels and at five passes for titanium and its alloys.

The interaction of the plasma channel with the surface of the electrode workpiece is not always purely thermal; nonetheless the surface of the piece is often enriched with elements contained in the environment and with the contents of the tool-anode material. The penetration depth [13] of these elements in the superficial layer of the piece is a function of both the energy of the discharge pulse and the size of the gap and can be expressed with the relation:

$$h = \frac{kW_S}{AS} \qquad (3)$$

where $W_S = \int_o^\tau U(t)I(t)dt$ is the energy emitted in the gap during a singular discharge; U and I are the voltage and the current in the gap, respectively; τ is the pulse duration; A is the area of the workpiece surface processed at a singular discharge; and k is a constant that depends on the thermo-physical properties of the workpiece material.

2.2.1 Technology of powder deposition formation

From the results of the experimental researches carried out by the authors [12, 13], it has been demonstrated that, in order to obtain the powder depositions, it is necessary to introduce the powder in the anode zone of the gap. From a technological point of view, a parameter that needs to be predicted is the thickness of the deposition layer. This parameter is quite important, because in some cases it is determinant in the technological application of the method. Analysing the experimental results of the authors of this chapter and those of the works [13, 18, 19, 29, 30], in the case of the deposition layers of different materials, the following relationship was deduced for the deposition thickness:

$$H = \frac{P^c \cdot f^d \cdot W^k \left(a - bS^2\right) \cdot r^m}{\rho \cdot A} \cdot n \tag{4}$$

where P is the powder flow, f is the frequency of impulse discharges, r is the equivalent radius of powder particles, W is the energy released in the gap, S is the size of the gap, ρ is the density of the particles' material, A is the area of the processed surface, n is the number of passes of the tool-electrode on the workpiece surface, a and b are individual constants of the deposition materials, and c, d, k, and m are the power exponents that are experimentally established and constitute both functions, of the properties of the powder material and of the processing conditions.

Taking into account the experimental results obtained by the authors of the papers [13, 21] for the determination of the thickness of the deposition layer, the relation (Eq. 4) can be written as

$$H = \frac{\Delta m}{\rho \cdot A} \cdot n \tag{5}$$

where

$$\Delta m = P^c \cdot f^d \cdot W^k \left(a - bS^2\right) r^m \tag{6}$$

The uniformity of deposition of the layer is characterized by the thickness of the deposited layer (or by the dimension Δm on each surface unit). If the processing time is different during the manual processing on different surface portions, the surface is not continuous; the mechanization of the process ensures the formation of uniform deposition on the entire surface of the workpiece.

The compactness of the layers and the existence of recesses or asperities are visually appreciated or measured using the microscope analysis method.

2.2.2 Technology of oxide films formation

In order to achieve the experimental researches, the pulse current generator with priming from a high-voltage block with 12 kV voltage and 0.3 μA current was used; the laboratory installation ensures the positioning, fixing and rotation of the samples during the processing with an adjustable rotation frequency ranging from 0 to 150 rpm [28]. The used tool-electrode was made in the form of cylindrical bar, rounded at the working end in the shape of a hemisphere and made of stainless steel.

The cathode piece was a cylindrical bar 13 mm in diameter. The piece was firmly fixed to the tool-machine frame and the tool-electrode into the tool port. The workpiece was connected to the pulse generator as cathode and the tool-electrode as anode, their axes forming an angle of 90°. In the case of flat surface processing, both the piece and the tool-electrode were made in the form of cylinders with a diameter of 10 mm; they were firmly fixed to the tool-machine port with parallel active surfaces and the electric discharge scanning the workpiece surface by migration of the plasma channel on these surfaces at a processing cycle. The materials for the pairs of electrodes (workpiece/tool-electrode) were chosen: aluminium alloy $AlCu_4Mg_1$, steel C_{45}, titanium alloy $TiAl_6Mo_4$ and copper alloys.

2.2.3 Technology of graphite deposition formation

Experimental research on graphite deposits was carried out under normal conditions in the air environment, under sub-excitation regime of PEDM. For the purpose of realizing the experiments, the power supply having the following parameters was used: the energy released in the gap W_S = 0–4.8 J, the energy accumulated on the capacitor battery W_c = 0–12 J, the voltage on the capacitor battery U_c = 0–200 V, the capacity C = 100–600 μF with step of 100 μF, the gap value S = 0.05–2.5 mm, discharge frequency f = 0–50 Hz and impulse duration τ = 0–250 μs. Due to these parameters, we can provide the PEDM operation under the regime of "hot" electrode spots (with the melting of the processed surfaces) and under the regime of the "cold" electrode spots (without melting the surfaces undergoing processing, although at nanometer scale it occurs).

Between the two electrodes, a graphite cathode and an anode made of the metal specimen, microelectric discharge is applied. The microelectric arc that is produced for a very short period of time—microseconds—has a very high temperature of about 10^4°C. At this temperature the graphite erodes (in the form of separated carbon atoms or chemical compounds of the type CO and CO_2 which further decompose into carbon and oxygen, the first being ionized is deposited in the form of film on the metal surface, and the oxygen is released in the plasma [28]), and due to the fact that it is in an electric field, the graphite particles in the vapour state are transported to the opposite sign electrode (anode) made up of the metal specimen. Thus, on the metal surface of the anode, a graphite layer of the micrometre dimension is deposited.

2.3 Methodology for determining the corrosion potential and speed of the samples

In order to measure the potential and the speed of corrosion, a setup was developed, the scheme of which is shown in **Figure 1**. In vessel 1 with electrolyte 2 (3% NaCl water solution for samples made of construction steels and 40% H_2SO_4 water solution for titanium, copper and aluminium alloy samples), the sample and cathode 6 are fixed. The cathode joins at the "−" pole and the sample at the "+" pole of the DC source with smooth voltage regulation. A milliammeter is connected in serial with the source and a voltmeter, in parallel, respectively. To measure the corrosion potential of the superficial oxide layer 4, the base metal 3 is placed in a rubber gasket 5 having a hole 7 so that only the oxidized surface of the sample is in contact with the electrolyte. The voltage is gradually increasing until the current appears, which will be indicated by the voltmeter—the value of the corrosion potential and the milliammeter—the corrosion current. In order to increase the measurement precision, several measurements are made for different parts of the sample surface.

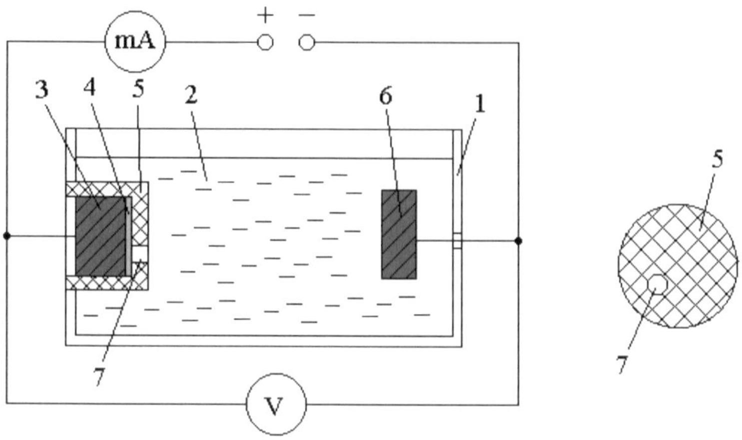

Figure 1.
Potential and speed of corrosion measurement scheme: (1) vessel, (2) electrolyte, (3) sample connected as anode, (4) oxide layer, (5) rubber gasket, (6) cathode and (7) hole for contacting the oxidized surface with the electrolyte in the water solution.

In order to perform the chemical corrosion resistance tests of the processed samples (with PEDM-coated oxide films on the active surface), electrolytes from distilled water and chemical agents with the respective concentration at room temperature were prepared, or, in separate cases, the device was continuously fed with an electrolyte of a certain temperature via a thermostat.

The samples were firmly fixed in the prepared device (**Figure 1**) so that only a circle-shaped part of the processed surface was subjected to dissolution. This was included in the anode dissolution circuit as the anode. The tests were carried out in the operating mode of the experimental DC installation and the 0.1 V step-by-step voltage change [28].

The corrosion speed is determined by weighing the samples before and after the test at the analytical electronic balance KERN ABS-N ABJ-NM with the accuracy of 10^{-5} g.

The corrosion speed index is determined with the relation (Eq. 3):

$$K = \frac{\Delta m}{S \cdot t}; \tag{7}$$

where K is the speed of corrosion (mass indicator), $g{\cdot}m^{-2}{\cdot}h^{-1}$; S is the area of the workpiece surface, m^2; t is the test duration (work time), h; and Δm is the mass lost (or the mass addition), g:

$$\Delta m = m_i - m_f \tag{8}$$

where m_i is the initial mass of the sample, g, and m_f is the final mass of the sample, g.

3. Results

3.1 Research on structural and phase transformations in superficial layers of the workpiece

The formation of powder deposition layers is accompanied by a wide range of structural and phase changes [13, 16, 29, 30]. The phase transformations are conditioned by the occurrence of the liquid phase and the vapour of the deposited particles interacting with the discharge channel, the environment, the cathode material and the liquid phase and solid phase transfer and hardening processes. As it has been shown, due to the electroerosion processes of the cathode material, it may be in the liquid-vapour state and interact with the environment; therefore, usually in the formed powder layer, besides oxides, nitrides and its intermetallic components, cathode material can be introduced in its other phases, respectively [13, 16, 29, 30]. The mentioned phases may occur in the contact zone between the particle and base material during the interaction of the liquid-vapour material of the cathode with the gap medium and the deposited particle and also in the case of deposition of precursor particles, which do not come into direct contact with the material of the piece but which have been influenced by plasma discharge. In particular, this is the area of "cold" spots, whose dimension is larger than the contact area between the sample and the particle (even for the maximum particle diameter of the deposited powder), and this is clearly visible for $S > 0.5$ mm (see **Figure 2**).

The condition in which the deposited particles interact with the surface portion of the workpiece, which has previously been in contact with the plasma discharge, is more often achieved in obtaining the continuous deposition layers.

These conditions were achieved in the roentgen research of the phase component of superficial layers for steels (steel 3, steel C45) and titanium (BT 1-0). Plasma interaction was achieved in the continuous processing (without including the powder in the processing area) when the interaction areas partially or variously overlapped (the number of passes n on the same surface portion varied). The cylindrical electrode-anode was made of graphite. The gap value was changed in the range of 0.15–2 mm, and the other parameters were Uc = 240 V and f = 40 Hz.

The analysis of the change of the phase composition of the processed surfaces (steel C45, titanium BT 1-0) has shown that the basic role in these changes is played by rapid heating and cooling, and their interaction with the environment, during the processing, is not so significant. The intensity of these changes is a function of the deposition material properties and processing conditions, such as the gap value (which, during the experiments, was modified for constant energy emitted in the

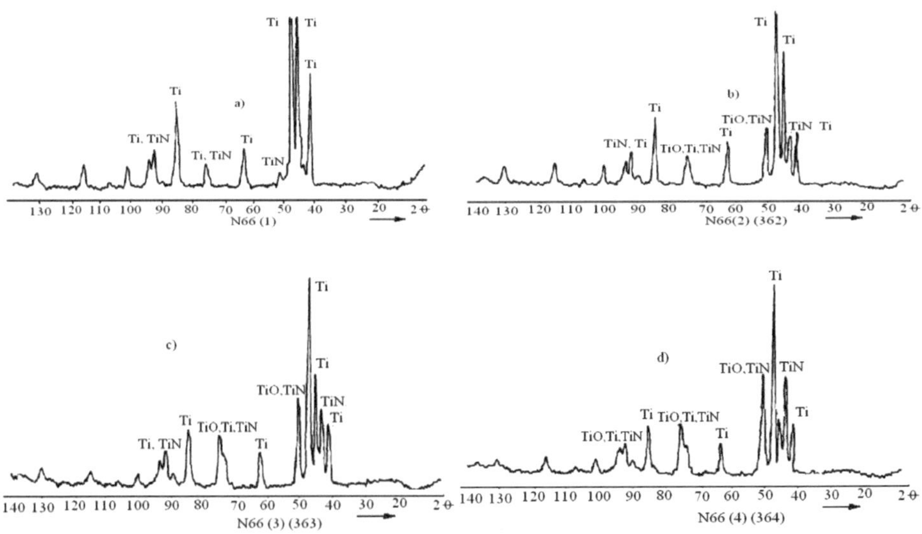

Figure 2.
Diffractograms of titanium BT 1-0 cathode sample surfaces processed by PEDM at a different number of passes, n: (a) 2 passes, (b) 4 passes, (c) 6 passes and (d) 8 passes; the energy emitted in the gap W = 5,3 J; the gap value S = 1.5 mm.

gap W = const). While processing BT-1-0 technical titanium, practically all the phases that are possible in its interaction with the oxygen and nitrogen in the air, under the action of discharge, have been experimented (**Figure 2**) [13]. Oxides (TiO, Ti_2O) and titanium nitride (TiN) form more intensively at gap values $S < 1$ mm. For larger gaps ($S \geq 2$ mm), only traces of TiO and TiN can be identified (**Figure 2**). In the multiple action of plasma, even for large values of the gap, the same phase component is formed on the surface of the titanium (**Figure 2b**), that is, TiO, TiO_2 and TiN. In this regard, less active is the steel surface; the increase of its processing time (on the account of the number of n passes) changes nonessentially its phase composition (**Figure 3**).

After the action of the discharge plasma on the surface of the steels on the diffractograms, along with the characteristic lines for the ferrite phase, there are also lines corresponding to the austenite phase. The maximum austenite quantity is obtained for $S \leq 0.5$ mm, i.e. when the outbreak surface is practically occupied by the liquid sample material. Some of the lines of the diffractograms can be identified as characteristic to carbide lines (Fe_3C) and to iron nitrides lines (Fe_3N-Fe_4N).

The presence of austenite and cementite lines during the processing of samples made of steel 3 is caused by the transfer on the surface of the anode material (graphite). The absence of oxides for the investigated regimes according to the roentgen method, at first estimation, under normal conditions, confirms the

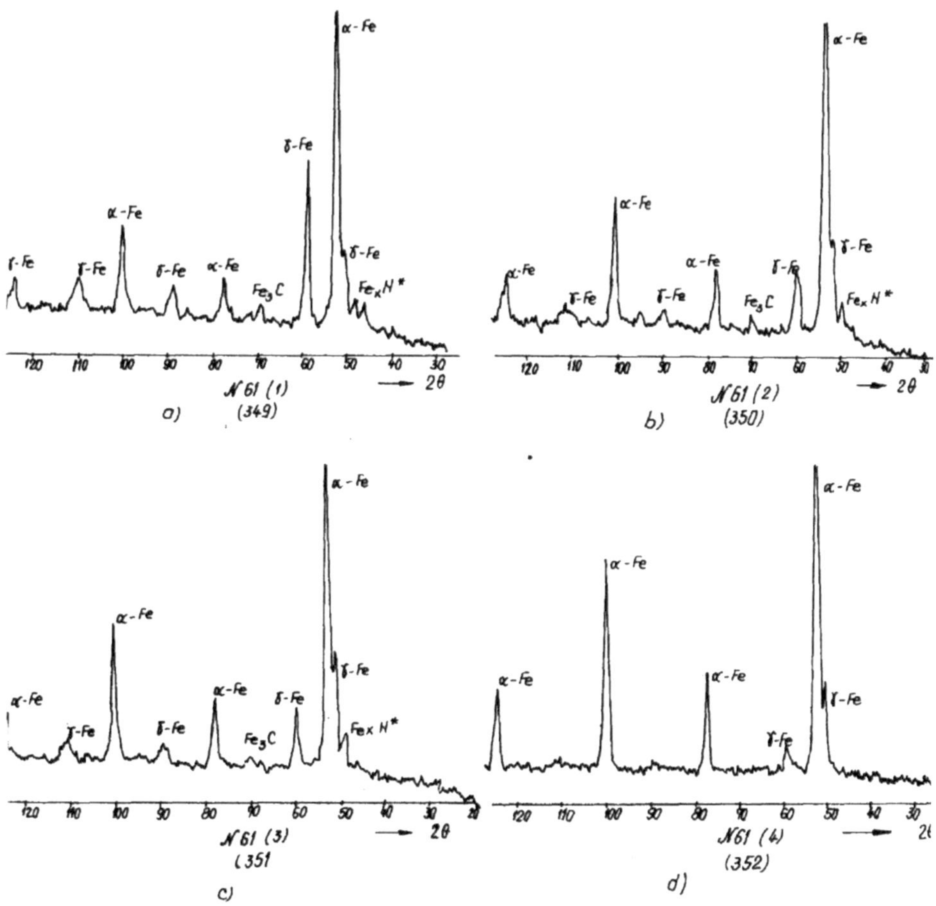

Figure 3.
Diffractograms of steel C45 cathode sample surfaces processed by PEDM for different gap values: S = 0.15 mm (a), S = 0.45 mm (b), S = 1.0 mm (c) and S = 2.0 mm (d).

hypothesis that the "cold" spots burn the oxides and impurities of the superficial layer [31–33], i.e. the discharge channel theoretically "migrates" to these defects of the surface, which finally condition their removal by vaporization (this is valid for singular passes) and also for the reduction of oxides by carbon. The technological process for steel surface cleaning by arc discharge in vacuum is developed based on this phenomenon [34].

The detailed investigations [13] of the processed surfaces using the Mössbauer method of the steel samples (**Figure 4**) indicate both spectra and represent an elusive superposition of iron oxides and hydroxide doublets and confirm the presence of the *γ-Fe* phase for $S = 0.5$ mm. The distribution of iron, oxygen, carbon and nitrogen in the sample depth is shown in **Figure 5**, which shows that carbon

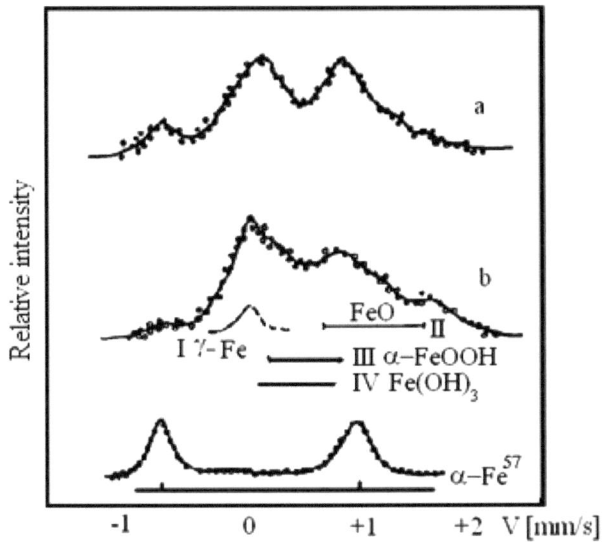

Figure 4.
Mössbauer spectra of steel 3 sample after processing by PEDM: S = 2 mm (a) and S = 0.5 mm (b) [13].

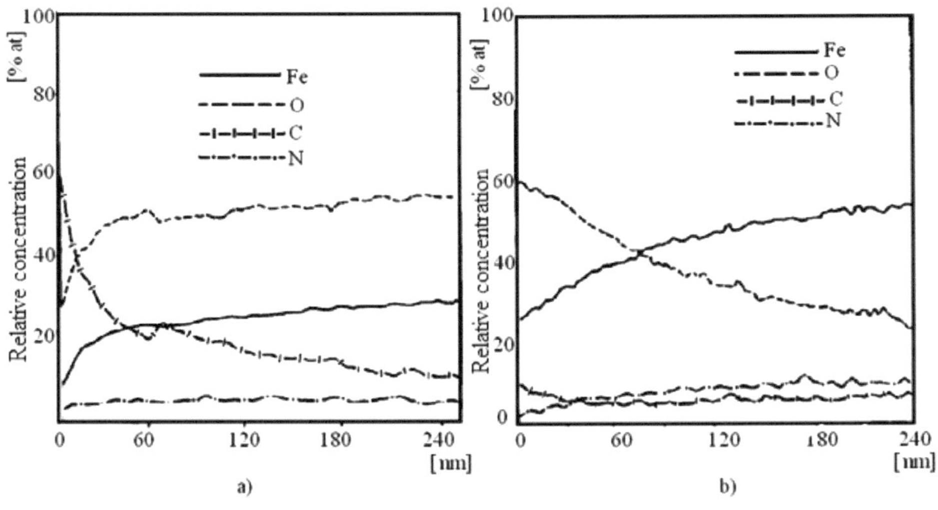

Figure 5.
Elements distribution in steel 3 sample surface after processing by PEDM for gap and specific energy values: $S = 0.5$ mm and $P = 3.19\ J{\cdot}mm^{-3}$ (a); $S = 2$ mm and $P = 1.5\ J{\cdot}mm^{-3}$ (b) [13].

aggregates to the surface. Its concentration decreases rapidly as the depth of the investigated layer increases, and the oxygen concentration reaches up to 60%.

This increased oxygen concentration can be achieved in the formation of hydroxides in the superficial layer of the sample. At a depth of layer of up to 240 nm, the nitrogen concentration is up to 5–6%.

For the value of gap $S = 2$ mm, the main components of the surface are iron and oxygen, and the secondary are carbon and nitrogen, respectively (**Figure 5b**). The oxygen concentration reaches 60% on the processed sample surface and lowers in the depth of the sample; the variation can be explained by the fact that in the superficial layer, hydroxides are formed and, in the depth of the sample, metastable oxides are formed; according to roentgenograms FeO (a single paramagnetic iron oxide at room temperature) is missing on the surface of the sample. Therefore, we can accept meta-stability of some very thin layers of Fe_2O_3 and Fe_3O, which is known according to [13]. Moreover, the possibility of amorphous iron hydroxides formation is not excluded [13]. Mössbauer spectra show that the complexity of unambiguously identifying amorphous hydroxides is very difficult.

The physical composition of the titanium and steel surfaces, after the removal of a layer of 5–10 nm thickness by means of rectification, changes nonessentially: the amount of oxides decreases in titanium and in steels—the nitride component decreases. Changing the energy regime does not condition the essential qualitative changes, for constant values of the gap S = const, and only quantitative changes take place.

Thus, the analysis of diffractograms has shown that among the elements of the processing regimes, the most important that essentially influence the phase component of the processed surface are the size of the gap, the energy density in it and the time of interaction with the plasma (the impulse duration and the number of passes of the plasma channel on the processed surface).

3.2 Determination of corrosion resistance of titanium and its alloys with deposited layers of metallic powder

In the process of deposition with electrodes made of compact materials by PEDM method, it is difficult to form layers with a full continuity, constant thickness, without pores, without impurities, etc., important elements for obtaining corrosion-resistant layers. The depositions obtained by the authors of the papers [13, 16, 29, 30] have shown good results in their corrosion resistance. When depositing metal powders, it is possible to form layers with highest thickness, but the problem of porosity remains unresolved.

Up to now, detailed research has been carried out on the corrosion resistance of titanium samples on which palladium (Pd), ruthenium (Ru) and nickel (Ni) depositions have been formed [29, 30]. The modification of the superficial titanium chemical composition with these metals was performed in order to obtain anodes on the base of titanium resistant to corrosion. The obtained results [29, 30] indicate that deposits formed by this method increase the corrosion resistance of titanium by about five times.

The corrosion and the electrochemical behaviour of titanium, with layers formed with the use of metal powders, were investigated by the authors [30] for the application of the 10% H_2SO_4 solution in distilled water at temperatures of 80 and 100°C, respectively. After the formation of the layers, some of the samples were subjected to annealing in vacuum at 1150°C for 1 hour for the purpose of palladium diffusion in the sample material. The corrosion and the electrochemical tests for the annealed samples were carried out in 10% H_2SO_4 solution in distilled water at 80 and 100°C and in 20, 30 and 40% H_2SO_4 solution in distilled water at 100°C. The palladium content was determined in the solution by the photo-calorimetric method. In the annealed samples, the palladium content was 5.6 mg·cm^{-2}.

The corrosion speed of titanium with powder depositions in 10% H_2SO_4 solution in distilled water at 80°C was 0.2 g·m^{-2}·h^{-1}; at 100°C in the first 5 hours of testing, the speed was 0.78 g·m^{-2}·h^{-1}; and then, during the tests, the speed decreased and, in 50 hours of testing, has dropped to 0.3 g·m^{-2}·h^{-1}. This situation of diminishing the corrosion speed can be caused by the dissolution of the unstable phases in the first hours. The corrosion potential, under these conditions, is situated within the passivity limits and equal to +400 mV in 10% H_2SO_4 solutions at 80°C (relative to the standard hydrogen electrode) and at 100°C equal to +300 mV compared to the titanium samples, without protecting depositions in the 10% H_2SO_4 solution in distilled water, which actively dissolves at potential of −0.56 mV with the speed of 18.72 g·m^{-2}·h^{-1}. Thus, we can state that, in the case of deposits application, the corrosion speed decreases up to 100 times.

After annealing, the corrosion speed has become even smaller.

For 10% H_2SO_4 solution in water at 80°C, this was 0.056 g·m^{-2}·h^{-1}, and at 100° C the corrosion process was completely absent.

With the increase in the solution concentration to 20, 30 and 40% H_2SO_4, respectively, in water, the corrosion rate at 100°C increased accordingly to 0.5, 0.9 and 1.8 g·m^{-2}·h^{-1}. The corrosion potentials, under these conditions, are situated within the passivity limits and are equal to 10% H_2SO_4 solution in water at 80 and 100°C after 5 hours of testing to +700 mV, in addition, corresponding to solutions containing 20, 30 and 40% H_2SO_4 in water; it ranged from +335 to +400 mV (**Figure 6**). For comparison, it can be mentioned that the corrosion speed of Pd layers in 10% H_2SO_4 solution in water at 100°C, depending on the specific processing time and regime, varies from 0.1 to 0.8 g m^{-2} h^{-1}. With increasing acid concentration in water, the speed increases for 40% H_2SO_4 solution at 100°C for different samples, ranging from 0.85 to 2.5 g m^{-2} h^{-1}.

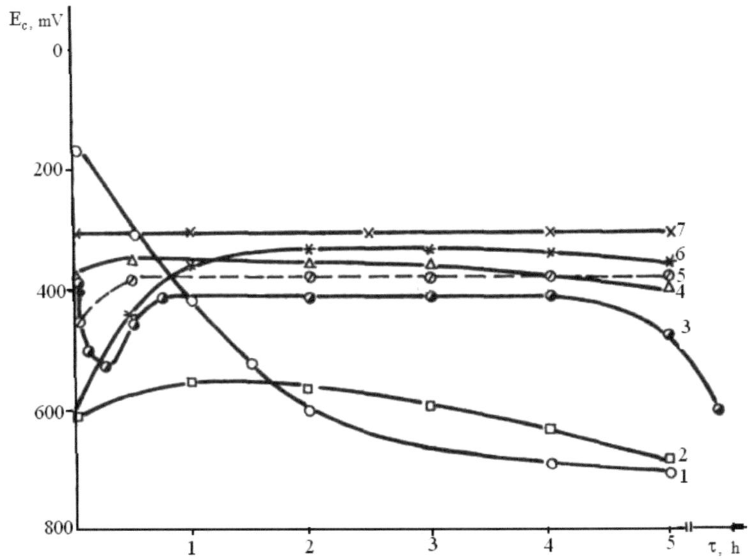

Figure 6.
Changing in time the corrosion potential of palladium-deposited samples (without annealing and after annealing) in water solutions of 10, 20, 30 and 40% H_2SO_4 at 80°C and at 100°C: 1–10% H_2SO_4, 80°C, without annealing; 2–10% H_2SO_4, 100°C, with annealing, 3–10% H_2SO_4, 80°C, without annealing; 4–40% H_2SO_4, 100°C, with annealing; 5–30% H_2SO_4, 100°C, with annealing; 6–20% H_2SO_4, 100°C, with annealing; and 7–10% H_2SO_4, 80°C, without annealing.

The research made by authors, in order to determine the corrosion resistance of titanium and its alloys with the deposited metallic powders, has shown that their corrosion speed can be reduced by 100 times in water solution containing 10–40% H_2SO_4 at the temperature of 100°C, and, in the case of additional subjecting, the annealing process and the corrosion speed can be reduced by more than 100 times, with the intention of homogenizing the structure of the layer and developing the diffusion processes of the elements deposited in the material of the piece. These layers allow to increase the corrosion potential up to +400 mV in 10% H_2SO_4 solution at 80°C and up to +300 mV at 100°C, while the titanium, in the absence of protective depositions, dissolves very actively at the potential of −0.56 mV.

Finally, we can conclude that the deposition of anticorrosive coatings on the surfaces of titanium and its alloy pieces can be applied in the construction of machines, ensuring a considerable duration of their functionality in aggressive environments.

3.3 The results of research on the corrosion properties of oxide films

In the researches carried out by the authors, the results of the experimental measurement of active electrical surface resistance for the samples made of C45 steel, processed by applying PEDM, under atmospheric conditions, at room temperature, are presented both for the surface of the anode-electrode samples and for the surface of cathode workpiece (**Table 1**).

The electrical surface resistance of the unprocessed samples by PEDM plasma ranges from 0.05 to 0.09 Ω mm^{-2}. From the analysis of the experimental results presented in **Table 1**, we can state that, in all cases, there is a substantial increase in the surface active resistance of the electrodes (both of anode and cathode) that participated in the PEDM process, but the active surface resistance of the anode-electrode sample is about three times smaller than that of the cathode-workpiece sample. The last finding can be explained by the fact that at the surface of the anode, under the same conditions, a higher amount of energy is released, and possibly more intense vaporization processes take place, and, on the other hand, the oxygen ions are predominantly directed by the forces of the electric field from the gap to the cathode surface, which is why the intensity of the oxide film formation is less intense on the anode surface.

Sample	Electrical surface resistance, $\times 10^6$ Ω mm^{-2}				
	Experimental data				Average value
Cathode	0.88	0.72	1.46	0.71	0.98
	0.97	1.52	0.68	0.72	
	0.73	1.09	0.76	0.83	
	1.33	1.10	1.04	0.73	
	1.07	0.88	1.21	0.78	
Anode	0.81	0.26	0.46	0.31	0.33
	0.11	0.14	0.56	0.34	
	0.29	0.11	0.62	0.38	
	0.87	0.38	0.11	0.15	
	0.12	0.13	0.27	0.17	

Table 1.
Electrical surface resistance of oxide films for steel C45 samples.

Processed sample material	Average value of surface resistance, $\times 10^6\ \Omega\ mm^{-2}$
Titanium alloy BT8 (TiAl6Mo4)	1.6
Duralumin Д16 (AlCu4Mg1)	0.25
Technically pure copper M0	0.15
Bronze БрА5 (Cu95Al5)	0.17
Brass Л63 (Cu63Zn37)	0.19

Table 2.
Electrical surface resistance of oxide films formed on the workpiece-cathode surfaces.

Further (**Table 2**), the measurements of the surface electrical resistance of the superficial layer for cathode samples made from titanium, aluminium and copper alloys were made.

As can be seen from the results presented in **Tables 1** and **2**, the electrical surface resistance is a function of the material of the investigated samples and has higher values for the materials with an increased oxygen avidity.

On the basis of the corrosion resistance tests, the volt-ampere characteristic was constructed both for PEDM-treated surface samples under air conditions and for samples with unprocessed surfaces (**Figure 7**).

If we compare the results presented in **Figure 8**, we can see that the anode dissolution potential of the processed sample has increased by six times compared to that of the unprocessed sample. As the voltage applied to the electrodes increases, the anode dissolution table of the investigated pieces changes. For example, in the case of voltage $U = 2$ V, the current in the circuit for the raw surface is 300 mA, and for the sample with oxide films, it is only 161 mA.

From the volt-ampere characteristic shown in **Figure 8**, we can see that the volt-ampere characteristics for the unprocessed and processed sample in the range 1.5–2.5 V are practically parallel, which shows the presence of an isolation film on the surface of the processed sample which increases its active resistance.

If we take into account the results obtained [28] for the case of measuring the surface active resistance, then we can conclude that the oxide layers formed on the sample surfaces do not have a total continuity, which is why the anodic dissolution process takes place more intensively than we expected.

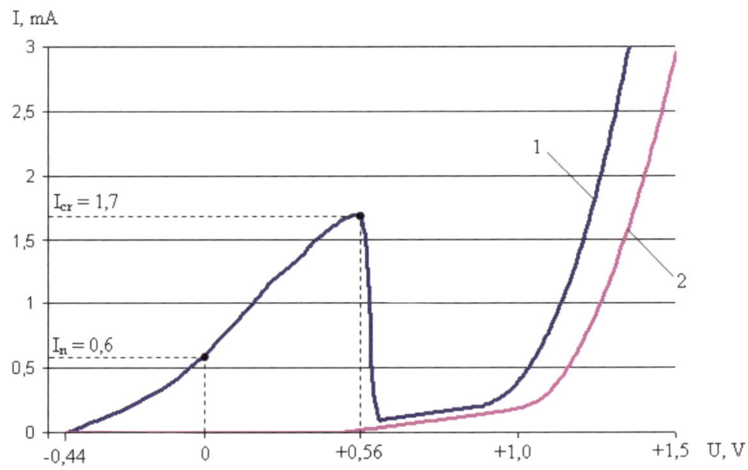

Figure 7.
The volt-ampere characteristic of the corrosion process of the steel C45 samples in the electrolyte (3% NaCl water solution): (1) the raw surface and (2) the processed surface.

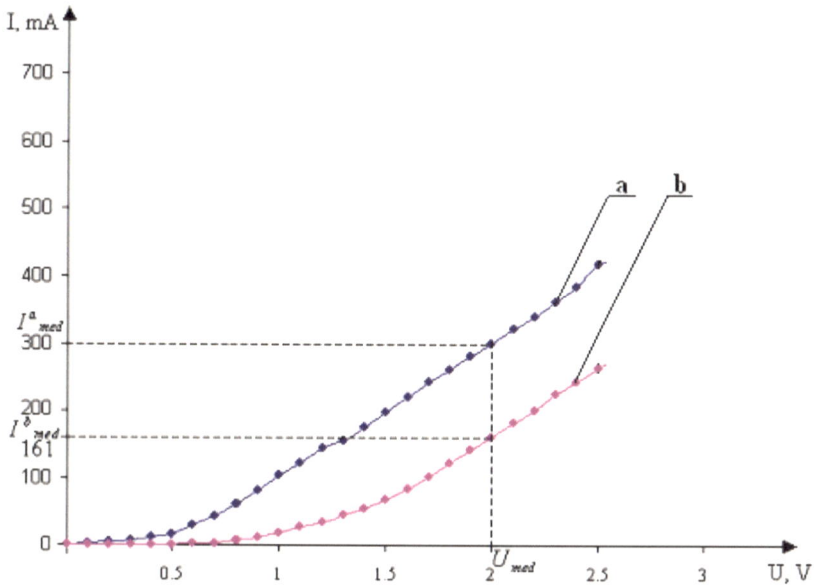

Figure 8.
The volt-ampere characteristic of the electrochemical corrosion process: (a) unprocessed surface and (b) processed surface.

No.	Initial mass of the sample m_i, g	Final mass of the sample m_f, g	Δm, g	Speed of corrosion K, g m^{-2} h^{-1}
1	12.8441	12.8237	−0.0204	812.10
2	13.1458	13.1254	−0.0204	812.10
3	13.3225	13.3025	−0.0200	796.17
4	14.0610	14.0417	−0.0193	768.31
5	13.0248	13.0031	−0.0217	863.85

Table 3.
Speed of anodic dissolution of steel C45 samples with unprocessed surfaces [28].

No.	Initial mass of the sample m_i, g	Final mass of the sample m_f, g	Δm, g	Speed of corrosion K, g m^{-2} h^{-1}
1	12.8569	12.8441	−0.0128	509.55
2	13.1605	13.1458	−0.0147	585.19
3	13.3338	13.3225	−0.0113	449.84
4	14.0717	14.0610	−0.0107	425.95
5	13.0366	13.0248	−0.0118	469.74

Table 4.
Speed of anodic dissolution of steel C45 samples with oxidized surfaces [28].

For the purpose of determining the corrosion speed, five pairs of unprocessed and processed by PEDM plasma samples were subjected to testing under the same anodic dissolution conditions. Based on the measurements, **Tables 3–6** were completed.

If we analyse the obtained results, we can see that, for the case of processed samples made of C45 steel, the corrosion speed in 3% NaCl water solution, at the applied voltage of 2 V, is, practically in all cases, twice lower than for the

No.	Initial mass of the sample m_i, g	Final mass of the sample m_f, g	Δm, g	Speed of corrosion K, g m^{-2} h^{-1}
1	1.6803	1.6700	−0.0103	103
2	1.6700	1.6599	−0.0101	101
3	1.6599	1.6466	−0.0133	133
4	1.6466	1.6324	−0.0142	140
5	1.6324	1.6198	−0.0126	126

Table 5.
Speed of anodic dissolution of titanium BT8 samples with unprocessed surfaces.

No.	Initial mass of the sample m_i, g	Final mass of the sample m_f, g	Δm, g	Speed of corrosion K, g m^{-2} h^{-1}
1	1.9582	1.9578	−0.0004	4
2	1.6615	1.6613	−0.0002	2
3	1.8829	1.8828	−0.0001	1
4	1.8325	1.8324	−0.0001	1
5	1.7648	1.7647	−0.0001	1

Table 6.
Speed of anodic dissolution of titanium BT8 samples with oxidized surfaces.

unprocessed samples by oxidation. The speed of corrosion of titanium samples with oxide films in 30% H_2SO_4 water solution at 80°C decreases by about 100 times comparatively with the unprocessed titanium surfaces.

If it is taken into account that the potential for natural corrosion is tens and hundreds of times less than in the case of the tests carried out, then the effectiveness of the application of oxides and hydroxides, in the amorphous state, will increase more significantly. For example, if we return to **Figure 7**, then we can see that, for the voltages applied in the limits of 0 and +0.4 V, the current in the circuit, formed by the workpiece, does not exist; consequently, the corrosion speed will be zero, whereas, under the same conditions, for the raw piece, there are considerable currents; therefore, the corrosion speed is significant. Finally, we can admit that the application of amorphous oxides and hydroxides to metal surfaces by PEDM is beneficial for increasing their active corrosion resistance.

3.4 The corrosion resistance of the graphite film

In order to determine the corrosion resistance of the graphite films, the method of electrochemical research was chosen. This method was used because it allows the acceleration of the corrosion process, which is nothing more than anodic dissolution, and, thus, it shortens the study of a sample [28].

The electrochemical installation is composed of an electrolysis cell containing a 1% NaCl water solution in which two electrodes are placed, a steel cathode and anode, also made of steel or graphite film-coated steel. They are connected by a voltmeter and an ampere metre at an adjustable DC source in the range of 0–40 V [28].

In order to determine the value for the potential of the electrochemical dissolution, it is necessary to build the volt-ampere characteristic, for the sample anode steel C45 and for the sample anode steel C45 covered with graphite. The cathode material, in both cases, is steel C45.

For the beginning, the samples were polished and rinsed with distilled water, then placed in the electrolyte solution at a distance of 10 mm one from another. A DC voltage, with an 0.4 V increments, is applied at a holding time of 3 min (**Figure 9**). The values for the electric currents are recorded by the ammeter, and they are presented in **Figure 9**.

When samples are placed in electrolyte, in the absence of external electric current, a stationary potential is established.

In the case when both electrodes are made from the same type of steel, the stationary potential is almost zero, and, in the case when the anode is covered with graphite, the stationary potential is 0.1 V.

When an external potential is applied, curves are obtained (**Figure 9** (curve 2)). It is observed that, for the steel sample, with the increasing potential, an increase in current intensity occurs as well, up to the value of 2 V, followed by an area where a decrease of intensity occurs up to the value of 2.9 V.

This effect may be due to the oxidation and hydroxylation chemical reactions that lead to the steel surface passivation, in the absence of the graphite film.

After this value, as shown in the chart, with the increase in the potential difference, the corrosion current decreases, and the corrosion process takes place at low speed, i.e., passive film emerges, which does not allow anodic dissolution.

The passive state is maintained until the potential of 2.8 V; then the current intensity increases considerably, accelerating the corrosion process.

An analogue behaviour is found while researching the graphite-coated steel sample, excluding that the passivation area appears almost at the value of 5.6 V.

Within the field of 5.6–6.2 V, the destruction of the deposited graphite film is observed, followed by an increase of corrosion currents.

The performance of the sample, covered with graphite film (**Figure 10**, curve 1), demonstrates that the presence of carbon significantly enhances the surface potential of the passivation ~3.2 V.

It is possible that the presence of graphite on metallic electrode surface generates deoxidation reactions, and the porosity of graphite film ensures the electrical contact between the electrolyte environment and the sample surface.

In **Figure 10**, the corrosion rate variation, depending on the time of immersion, is shown for the samples processed with graphite electrode tool.

From the analysis results, related to corrosion speed variation, depending on the duration of samples exposure in aggressive environment (**Figure 10**), we can

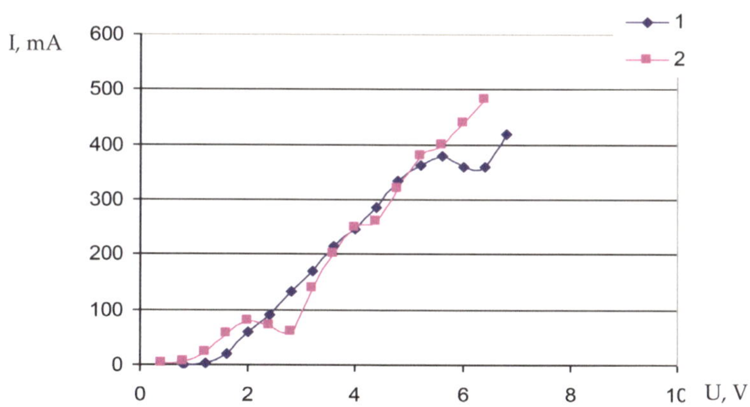

Figure 9.
Volt-ampere characteristic for the electrochemical process [28]: (1) steel C45 covered with graphite and (2) unprocessed steel C45 surface.

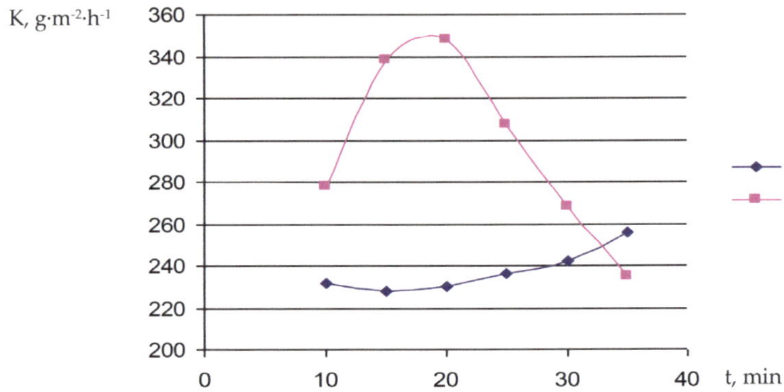

Figure 10.
Corrosion speed variation depending on the time of immersion [28]: (1) steel C45 covered with graphite and (2) unprocessed steel C45.

outline essentially higher corrosion rates for the initiation stage of corrosive process, in the case of unprocessed steel 45 without a graphite pellicle (**Figure 10**, curve 2).

Afterwards, with the passing of time, the process of corrosion is attenuated; the determined corrosion rate is smaller, when the time of immersion is longer.

Similarly, with the passing of time, the nature of the corrosion product is modified; oxides are most likely formed, having a low oxygen content and being more stable (Fe_3O_4) ($FeO{\cdot}Fe_2O_3$).

Simultaneously, it is acceptable that corrosion products form films on the metal surface that insulate the metal from corrosive agents. For this reason, the curves showing the variation of corrosion rate time show a trend of flattening.

These films are adherent and also compact enough so that the corrosion process diminishes in time, but does not annihilate.

It should be noted that the determinations were made in a static regime. In dynamic conditions, the formed oxide film can probably detach from the metal surface, and, in this case, the corrosion process is conducted with higher corrosion rates.

In the case of steel coated with graphite film, lower corrosion rates are found; afterwards, they begin to rise due to degradation of graphite film, as a result of its destruction.

Experimentation established that in acid solution of 30% HNO_3, after 3 min, the amount of uncovered graphite steel electrochemically dissolved 1.4 times greater than graphite-coated steel—1.3 times greater in 10 min. In salt solution of 1% NaCl, the amount of uncovered graphite steel electrochemically dissolved was 1.2–1.4 times greater than graphite-coated steel deposited by PEDM.

4. Conclusions

Analysing the results of the experimental researches, we can conclude the following:

- Thermal treatment, in the absence of liquid phase formation on the processed surface, can be performed only in the regime of maintenance of the electrical discharge in impulse on "cold" electrode spots.

- Under superficial processing conditions, with the maintenance of electric spell released on "cold" electrode spots, the mass transfer in the solid phase can reach depths of dozens of μm diameters, which will allow the working surfaces of the parts and cutting tools of the layers with prescribed physico-mechanical properties that will provide them with high-performance features.

- The thermal and chemical-thermal processes that result in the surface layers of the parts under the action of the plasma channel of the electrical discharges in impulse cause not only the modification of their chemical composition but also the modification of their physico-mechanical properties.

- Oxide films, formed as a result of interaction with PEDM plasma, increase the resistance by 10^6–10^7 times, which, in turn, influences the corrosion resistance.

Author details

Pavel Topala*, Alexandr Ojegov and Vitalie Besliu
Alecu Russo Balti State University, Balti, Republic of Moldova

*Address all correspondence to: pavel.topala@gmail.com

References

[1] Nanu A. Technology of Materials. Chisinau: Ştiinţa; 1992. 552 p

[2] Tian BR, Cheng YF. Electrochemical corrosion behaviour of X-65 steel in the simulated oil sand slurry. I: Effects of hydrodynamic condition. Corrosion Science. 2008;**50**(3):773-779

[3] LiuYing L, FuhuiWang L. Electrochemical corrosion behaviour of Nano crystalline materials: A review. Journal of Materials Science and Technology. 2010;**26**(1):1-14

[4] Rihan RO. Electrochemical corrosion behaviour of X52 and X60 steels in carbon dioxide containing saltwater solution. Materials Research. 2013;**16**:1 Epub Dec 04, 2012

[5] Kuphasuk C, Oshida Y, Andres CJ, Hovijitra ST, Barco MT, Brown DT. Electrochemical corrosion of titanium and titanium-based alloys. The Journal of Prosthetic Dentistry. 2001;**85**(2):195-202

[6] Huang C-L, Chen T-H, Chiang C-C, Lin S-Y. Electrochemical corrosion and mechanical properties of two biomedical titanium alloys. International Journal of Electrochemical Science. 2018;**13**:2779-2790. DOI: 10.20964/2018.03.26

[7] Song G-L. The grand challenges in electrochemical corrosion research. Frontiers in Materials. 2014. DOI: 10.3389/fmats.2014.00002

[8] Tian G et al. Investigation on electrochemical corrosion characteristic of 2A14 aluminum alloy in nitric acid. Surface Review and Letters. 2017;**24**(1850016):10. DOI: 10.1142/S0218625X18500166

[9] Topala P, Ojegov A. Formation of oxide thin pellicles by means of electric discharges in pulse. Annals of the Oradea University. Fascicle of Management and Technological Engineering. 2008. CD-ROM Edition. Edition of University from Oradea, Romania. ISSN 1583-0691, CNCSIS "Class B+";**VII**(XVII):1824-1829. DOI: 10.15660/AUOFMTE.2008.1213

[10] Negm NA, Salem MA, Badawi AM, Zaki MF. Corrosion inhibition properties of some novel n-methyl diethanolammonium bromide cationic surfactants. TESCE. 2004;**30**(2):853-869

[11] Gogoi PK, Barhai B. Corrosion inhibition of carbon steel in open recirculating cooling water system of petroleum refinery by a multi-component blend containing zinc(II) diethyldithiocarbamate. Indian Journal of Chemical Technology. 2010;**17**:291-295

[12] Topala P. Cercetări privind obţinerea straturilor de depunere din pulberi metalice cu aplicarea descărcărilor electrice în impuls. In: Rezumatul Tezei de Doctorat. Bucureşti: Universitatea Politehnica; 1993. p. 32

[13] Topală P. Cercetări privind obţinerea straturilor de depunere din pulberi metalice cu aplicarea descărcărilor electrice în impuls. In: Teza de Doctorat în tehnică. Bucureşti: Universitatea Politehnica; 1993. p. 161

[14] Abdulkareem MAA, Hani AA. The effect of zinc, tin, and Lead coating on corrosion protective effectiveness of steel reinforcement in concrete. American Journal of Scientific and Industrial Research. 2011;**2**(1):89-98. ISSN: 2153-649X. DOI: 10.5251/ajsir.2011.2.89.98

[15] Topală PA, Balcănuţă NP. Caracteristicile electrodinamice ale descărcărilor electrice în impuls. In: Tehnologii Moderne, Calitate, Restructurare. Chişinău: Universitatea Tehnică a Moldovei; 2001. pp. 203-208

[16] Topala P. Conditions of thermic and thermo-chemical superficial treatment innards with the adhibition of electric discharge in impulses. In: Nonconventional Technologies Review. Bucureşti: Editura BREN; 2005. pp. 27-30

[17] Topală P. Aplicări ale electroeroziunii în dezvoltarea tehnologiilor fine de prelucrare superficială a pieselor. Analele Ştiinţifice ale Universităţii de Stat “Alecu Russo” din Bălţi.—Serie nouă. Fasc. A. 2004. Tom XX, p. 66-69

[18] Топала ПА, Гитлевич АЕ, Корниенко ЛП. Коррозионное поведение титана с электроискровыми покрытиями. Москва: Защита металлов, 1989, vol. 29, № 3, pp. 351-356

[19] Топала Пи др. Влияние отжига на коррозионное поведение титана с электроискровыми покрытиями. Москва: Защита металлов. 1990. vol. 26. № 3. pp. 433-437

[20] Топала П и др. Упрочнение металлических поверхностей на участках для электроискрового легирования. In: Машиностроение и техносфера Xxi века. Vol. 3. Донецк: Сборник трудов XIII Международной научно-технической конференции; 2006. pp. 262-265

[21] Topală P, Stoicev P. Tehnologii de Prelucrare a Materialelor Conductibile cu Aplicarea descărcărilor Electrice în Impuls. Chişinău: Editura Tehnica-Info; 2008. p. 265

[22] Парканский НЯ. Исследования процесса электроискрового нанесения покрытий из порошковых материалов в электрическом поле. Дис. канд. техн. наук. Киев: Институт проблем материаловедения АН УССР; 1979. p. 27

[23] Ливурдов ВИ и д. Структура и эксплуатационные свойства деталей с покрытиями, полученными электроискровым легированием порошковыми материалами. ЭОМ. 1980; (5):33-35

[24] Парканский НЯ, Гитлевич АЕ. Особенности поверхности слоев катода при электроискровом нанесении порошковых материалов. ЭОМ. 1981; (1):32-35

[25] Гитлевич АЕ и д. Электроискровое легирование металлических поверхностей. Кишинев: Штиинца; 1985. p. 196

[26] Немошкаленко ВВ и др. Особенности формирования поверхностных слоев при искровых разрядах. Киев: Металлофизика. 1990. vol. 12. № 3. pp. 132-133. ISSN 0204-3580

[27] Мицкевич МК и др. Электроэрозионная обработка металлов. Минск: Наука и техника. 1988. p. 216

[28] Tiginyanu I, Topala P, Ursaki V, editors. Nanostructures and Thin Films for Multifunctional Applications. NanoScience and Technology. Switzerland: Springer International Publishing; 2016. p. 576 p. ISSN 1434-4904, ISBN 978-3-319-30197-6. DOI: 10.1007/978-3-319-30198-3

[29] Корниенко ЛП и др. Электрохимическое и коррозионное поведение титана с электроискровыми покрытиями Pd и Cr-Pd. Москва: Защита металлов. 1991. vol. 29. № 3. pp. 351-358

[30] Chernova GP et al. The influence of annealing on the corrosive behaviour of titanium with electrospark palladium surfaces. Metal Protection. 1990;**26**(3): 433-437

[31] Любимов ГА, Раховский ВИ. Катодное пятно вакуумной дуги. УФН. 1978;**125**(4):665-706

[32] Бушик АИ. Исследование динамики процессов при импульсном разряде на сложных электродах. Автореф. дис.

канд, ф.-м. наук. Минск: ФТИ АН БССР. 1973. р. 23

[33] Антонов СА и д. Дуговая эрозия катодов, содержащих включения эмиссионноактивной фазы. ЖТФ. 1982; **52**(52):266-270

[34] Булат ВЕ, Эстерлив МХ. Очистка металлических изделий от окалины, окисной плёнки и загрязнений электродуговым разрядом в вакууме. ФХОМ. 1987;(3):49-53

Section 2

Theoretical Studies

Chapter 7

Investigation of Corrosion Inhibitors Adsorption on Metals Using Density Functional Theory and Molecular Dynamics Simulation

Ambrish Singh, Kashif R. Ansari, Mumtaz A. Quraishi and Yuanhua Lin

Abstract

The use of computational chemistry as a tool in the design and development of organic corrosion inhibitors has been greatly enhanced by the development of density functional theory (DFT) and Molecular dynamic simulation (MD). Experimentally corrosion inhibitor development requires lots of money and time. Thus, in the era of hardware and software development, corrosion scientist can select a potential inhibitor on the basis of theoretical analysis of molecular properties of inhibitor molecules, which have reduced the cost. DFT and MD are capable to accurately predict the inhibition characteristics of inhibitor molecules using molecular/electronic properties and reactivity indices. The purpose of this book chapter is to summarize some important features related to DFT and MD, giving a brief background to the selected DFT/MD-based chemical reactivity concepts, calculations and their applications to organic corrosion inhibitor design. The impact of this book chapter is to illustrate the enormous power of DFT and MD.

Keywords: corrosion, DFT, MD, inhibitor

1. Introduction

Corrosion of active metals such as iron (Fe), copper (Cu) and Aluminum (Al) can be reduced by modifying their surface using organic/polymeric corrosion inhibitors. The inhibitor molecules adsorbed onto the metal surface and form a protective barrier against corrosion [1]. The corrosion protection ability of inhibitors depends upon the extent of interaction between inhibitor molecules and metal surface [2]. The basics mechanism through which adsorbed inhibitor molecules alter the corrosion process consists of either changing the cathodic and anodic corrosion reactions or physically blocking the active sites present on metal surface or by [3].

Perusal of the literature clearly reveals that huge number of publication exits where experimental works have been carried out to understand the corrosion inhibition process, however use of theoretical studies including density functional theory (DFT) and molecular dynamic simulation (MD) appeared in recent years. In traditional experimental work scientists has to test huge number of organic compounds for selecting them as a potential corrosion inhibitor. This kind of search takes more time and money. However, in the era of development in the field of hardware and software technology make us capable to select organic compounds as potential corrosion inhibitor without expensing huge amount of money on chemicals. The corrosion inhibition efficiency of organic compounds based on molecular and electronic properties can be predicted accurately using MD and DFT calculations. In the recent years, scientist have devoted their positive efforts in the field of corrosion using theoretical calculation [4–6]. To our knowledge, the review exists in the literature only describing separately DFT and MD and not both in relation to corrosion inhibition study [7, 8]. Here we are trying to describe both DFT and MD on the same platform simultaneously.

In this chapter we will try to explore some recent studies and concept which can provide some new concept in the field of computational chemistry related to corrosion inhibition phenomenon. Here, we summarize the adsorption of organic molecules onto the metal surface especially iron, aluminum and copper in order to study their corrosion inhibition properties using DFT and MD simulations.

2. Basic concepts related to molecular modeling

In literature, large number of publication/papers exits where corrosion inhibition property of inhibitors has been studied using theoretical study. In general organic molecules have the ability to donate electrons to empty d-orbital of metal surface and form covalent bond which is known as coordinate bond. However, they also undergo back donation, where they accept the free electrons presenting in filled metal orbitals and this is called as retrodonating bonds. Thus, these kinds of bonds make the organic compounds a potential candidate for corrosion inhibition. The theoretical parameters like highest occupied molecular orbital (HOMO), lowest unoccupied molecular orbital (LUMO), dipole moment (μ), energy gap (ΔE), Fukui indices, charge density, polarizability, softness, etc. [9–11] were discussed in DFT calculation.

DFT study separately represents the organic compound and metal surface [12–16]. In DFT, the interaction between the inhibitor molecule and metal surface was not directly modeled. However, in MD the most reactive fragment of the molecule that has greater affinity to interact with the metal surface is study using ab initio method. This method provide a real picture which is actually going in the corrosion process like interaction between inhibitor molecule and metal surface, orientation of molecule towards metal and organization of molecules.

3. Inhibitor/surface interaction study using DFT

3.1 The basics of DFT: the Hohenberg-Kohn theorem

Quantum electrochemistry is the field which includes quantum mechanics, electrochemistry and electrodynamics. In general, quantum electrochemistry is an application of density functional theory (DFT) for studying the electrochemical processes like transfer of electron towards electrode surfaces [17].

There are so many books and articles that exist in literature which describe about the basis set used in DFT [18–31]. So, here we provide only very few pints of DFT without using mathematics.

3.2 Functional

The functional is a function of a function, i.e., electron. Therefore, on the basis of fundamental quantum mechanics and from the experimental results different functional have been developed and are listed in **Table 1** [29].

There are several generations of density functional. First generation is known as X_α method and it is the simplest one. This functional was developed by J.C. Slater, who was working on the approximation of Hartree-Fock but unwillingly discovered the simplest for DFT. In $X_\alpha$functional electron exchange but not the correlation was included. The results obtained from $X_\alpha$method is as accurate as HF but in some cases it is better than HF.

The functional of second generation uses both the density and its gradients. In 1986 first gradient-corrected energy functional was proposed in which Becke, Perdew and Wang proposed exchange and Perdew correlation [32, 33]. In the present time the most popular functionalis of Becke and Perdew for exchange and correlation respectively [32], also Lee et al. [34] for correlation and Perdew and Wang for exchange and correlation both [35]. These functional are collectively known as "generalized gradient approximations (GGAs)."

The third generation is hybrid functional and it is more advanced than GGA. In this, the functional consists of both Hartree-Fock type exchange and DFT exchange calculated from the orbitals. In 1998, a hybrid functional called as B3LYP functional was introduced. In 1994, Gaussian introduced this functional in in computational package for first time and it is written by below equation:

$$E_{XC} = a_o E_{X(HF)} + a_1 E_{X(LSD)+a_2} E_{X(GGA)} + a_3 E_C \quad (1)$$

where $E_{X(HF)}$ represent the Hartree-Fock exchange energy, $E_{X(LSD)}$ stands for Dirac exchange energy and $E_{X(GGA)}$ and E_C are the gradient corrections to exchange energy, i.e., Be88form and Lee-Yang-Parr (LYP) functional respectively [34, 36].

3.3 Basis sets

In DFT calculation various basis sets are used [37]. The simplest one is STO-3G. In this basis set, 1s is given by three Gaussians and 2s, 2px, 2py, and 2pz each by

Acronyms	Name	Type
X_α	X alpha	Exchange only
HFS	Hartree-Fock Slater	HF with LDA exchange
VWN	Vosko, Wilks, and Nusair	LDA
BLYP	Becke correlation functional with Lee, Yang, Parr exchange	Gradient-corrected
B3LYP	Becke 3 term with Lee, Yang and Parr	Hybrid exchange
PW91	Perdew and Wang 1991	Gradient-corrected
G96	Gill 1996	Exchange
P86	Perdew 1986	Gradient-corrected
B96	Becke 1996	Gradient corrected
B3P86	Becke exchange, Perdew correlation	Hybrid
B3PW91	Becke exchange, Perdew and Wang	Hybrid correlation

Table 1.
List of density functional [29].

another three. Improvement in the basis set can obtain by incorporating two 1s functions for hydrogen and for 2nd row atoms like carbon two 2s and two 2p functions has incorporated. These functional are called as split-valence basis sets. Here valence and core orbitals are represented by two sets of functions and single set of functions respectively. An example for this:

Carbon 3-21G: Three Gaussians for 1s combination, same two Gaussians for 2s and 2p combination; plus 2s' and 2p' the same one Gaussian.

Carbon 6-31G: Six Gaussians for 1s combination, same three Gaussians for 2s and 2p combination; plus 2s' and 2p' the same one Gaussian.

In general it can be represented as *i-jk*, here i denote the number of Gaussians that are representing each core basis function, and *j* and *k* represent the numbers of Gaussians for split-valence basis functions.

Further advancement generates triple-split-valence basis sets, for example 6-311G. In this basis set core functional consists of six Gaussians and three sets of valence functions containing three, one, and one Gaussians, respectively.

Multiple zeta basis sets such as double zeta (DZ) and triple zeta (TZ) basis sets has also introduced. The difference between multiple and split-valence basis sets is that in multiple basis sets all of the orbitals are separated into either two or three sets of functions, and so forth. Also, multiple basis sets used different α-coefficients for s and p orbitals.

So, improvement in the electronic structure calculation can be achieved by adding functions corresponding to orbitals with a higher angular momentum. This concept

Basis	**Options**	**Atoms**
STO-3G	*	H–Xe
3-21G	* **	H–Cl
4-21G	* **	
4-31G	* **	H–Ne
6-21G	* **	
6-31G	+ ++ * **	H–Cl
LP-31G	* **	
LP-41G	* **	
6-311G	+ ++ * **	H–Ar
MC-311G	None	H–Ar
D95	+ ++ * **	H–Cl
D95V	+ ++ * **	H–Ne
SEC	+ ++ * **	H–Cl
CEP-4G	+ ++ * **	H–Cl
CEP-31G	+ ++ * **	H–Cl
CEP-121G	+ ++ * **	H–Cl
LANLIMB (except lanthanides)	None	H–Bi
LANLIDZ (except lanthanides)	None	H–Bi

Table 2.
Common basis set.

can explained as follows: for hydrogen atom, p functions are added; for carbon, nitrogen, oxygen, etc., d functions are added; and for transition metals, f functions are introduced. These are represented by adding an asterisk to the basis set and also specifying the p, d, f functions. This can be understood as per below example:

6-31G* or 6-31G(d): Adds d functions to 2nd row elements (C, N, O, etc.).

6.31G** or 6-31G(d,p): Adds d function to 2nd row elements (C, N, O, etc.) and p functions to H.

The lone pair of electrons on the heteroatoms can be taken into confederation by adding diffuse functions and these are represented in the basis set using + and ++ signs. For hydrogen atom polarization and diffuse functions are not very import but for other atoms like carbon, nitrogen, oxygen it is important, for example,

6-31+G: represents that diffuse functions has been added to 2nd row elements (C, N, O, etc.).

6-31++G: represents that diffuse functions has added to 2nd row elements and H.

In the case of transition elements which contain large number of electrons, requires more time for calculation. So, the LanL2DZ basis set is used for transition metal calculations. This basis set is derived by combining the valence electrons using double zeta functions and Los Alamos ECP. The software used in corrosion inhibition study with a wide range of basis sets are Spartan, Material Studio, Gaussian 03, Gaussian 09 and most recently Gaussian 14 has been introduced. **Table 2** represents commonly used basis sets [37].

4. DFT parameters and their application in corrosion inhibition study

4.1 Frontier molecular orbitals

The optical and electronic property of organic compounds is explained by analyzing their frontier molecular orbitals (FMO) like highest occupied molecular orbital (HOMO) and lowest unoccupied molecular orbital (LUMO) [38]. The interaction of organic compounds as corrosion inhibitor with the metal surface depends upon the electron donation ability of the organic compounds, i.e., ionization potential (*I.P.*). However, the electron accepting ability of organic compounds corresponds to the electron affinity (*E.A.*) [39]. So, HOMO is the electron donating orbitals and LUMO is electron accepting orbitals [40]. In general for a good corrosion inhibitor HOMO value should high, LUMO should low and energy gap between the LUMO and HOMO (ΔE) should low [40].

According to Koopmans' theorem the negative value of HOMO corresponds to the ionization potential (*I.P.*) and is given as follows [41–44]:

$$I.P. = -E_{HOMO} \tag{2}$$

Similarly negative value of LUMO corresponds to the electron affinity (*E.A.*) and is given below [41–48]:

$$E.A. = -E_{LUMO} \tag{3}$$

In literature Singh et al. [45] has correlated some important quantum chemical parameters of organic corrosion inhibitors using/B3LYP methods using 6-311G (d, p) basis set with their corrosion inhibition efficiency. They studied three triazine-based hydrazone derivatives and their molecular, orbital pictures and quantum chemical parameters are represented in **Figures 1–3**.

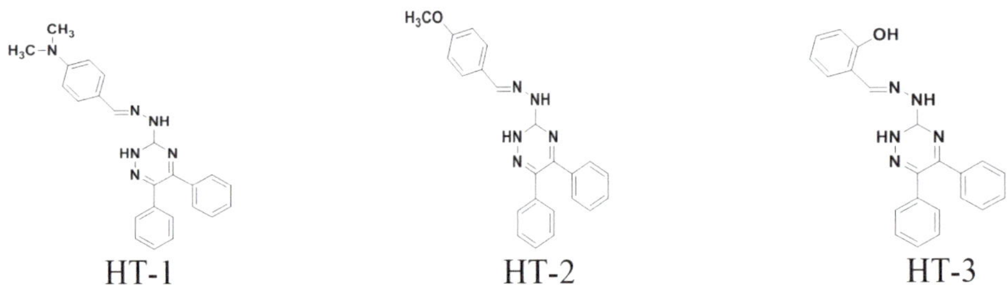

Figure 1.
Molecular structure of the triazine-based hydrazone derivatives. Adapted from Ref. [45].

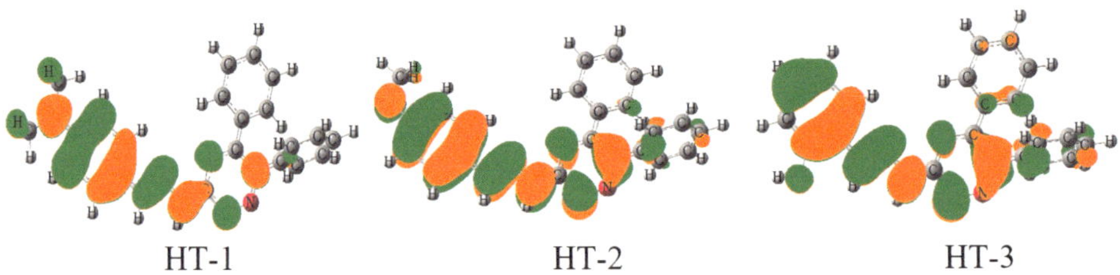

Figure 2.
HOMO orbitals of triazine-based hydrazone derivatives.

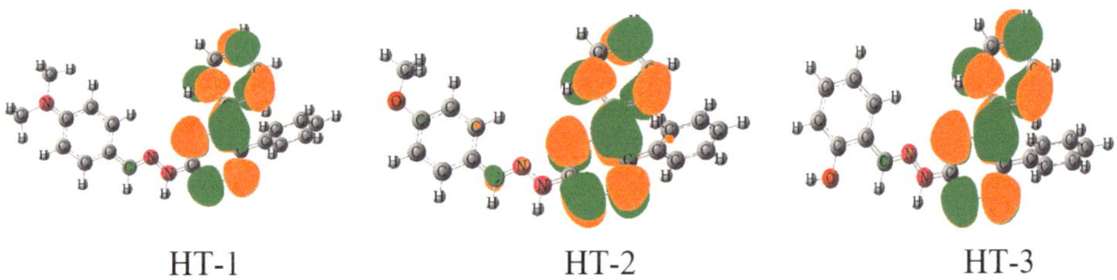

Figure 3.
LUMO orbitals of triazine-based hydrazone derivatives.

The HOMO and LUMO are represented in **Figures 2** and **3**.

From the figure, it could be said that HOMO regions are distributed on the triazine ring, phenyl ring containing electron donating group, nitrogen and oxygen atoms. These regions correspond to the electron donating areas and inhibitor molecules interact with the metal surface through these regions and undergo adsorption over it. However, the LUMO are distributed over the triazine and phenyl ring respectively and these are corresponding to the electron accepting regions. From these LUMO regions inhibitor accept electron from metal. Order of inhibition efficiency is HT-1 > HT-2 > HT-3.

4.2. Electronegativity and the electronic chemical potential

One of the important global molecular property calculated by density functional theory (DFT) which provide an idea for chemical reactivity and selectivity of organic compound as corrosion inhibitor are electronegativity (χ) and chemical potential (μ) [46]. Mathematical form of this is as follows [47, 48]:

$$\chi = -\mu = -\left(\frac{\partial E}{\partial N}\right)_{\nu(r)} \tag{4}$$

$$\chi = -\mu = -\left(\frac{I.P. + E.A.}{2}\right) \tag{5}$$

where $I.P.$ and $E.A.$ are ionization potential and electron affinity.

4.3 Global hardness and softness

According to Parr et al. global hardness (η) measures the resistance of an atom to transfer the charge [49] and it can be calculated using below equation:

$$\eta = \frac{1}{2}(E_L + E_H) \tag{6}$$

where E_L and E_H are the HOMO and LUMO energy.

The reciprocal of global hardness is global softness (σ) and given as [50]:

$$\sigma = \frac{1}{\eta} \tag{7}$$

4.4 Electric dipole polarizability

The electric dipole polarizability (α) is a measurement of the linear response of the electron density in the presence of an infinitesimal electric field F and it describes a second-order variation in energy:

$$\alpha = -\left(\frac{\partial^2 E}{\partial^2 F_a \partial^2 F_b}\right) a, b = x, y, z \tag{8}$$

The polarizability (α) is calculated as the mean value and express through above equation:

$$\alpha = \frac{1}{3}\left(\alpha_{xx} + \alpha_{yy} + \alpha_{zz}\right) \tag{9}$$

It was discovered that polarizabilities are inversely proportional to the third power of the hardness values [51–54].

4.5 Electrophilicity index

Parr et al. has introduced the concept of global electrophilicity index (ω) and is given as follows [55, 56]:

$$\omega = \frac{\mu^2}{4\eta} \tag{10}$$

The global electrophilicity index (ω) is also related to the ionization potential ($I.P.$) and electron affinity ($E.A.$) and is given as follows [56]:

$$\omega = \frac{(I.P. + E.A.)^2}{8(I.P. + E.A.)} \tag{11}$$

The electrophilicity index accounts the ability of the molecule to accept electrons. So, higher the ω value, greater would be the molecule to accept the electrons.

4.6 The fraction of electrons transferred

When the inhibitor and metal interaction occurs, flow of electron takes place from the lower electronegativity molecule to the higher electronegativity metal. This transfer of electron continues until the chemical potential becomes equal. The fraction of electron transfer is given as follows [57]:

$$\Delta N = \frac{\chi_m - \chi_i}{2(\eta_m + \eta_i)} \tag{12}$$

where metal and inhibitor molecule are represented by m and i indices respectively. The absolute electronegativity of metal and inhibitor molecule are represented by χ_m and χ_i respectively. The absolute hardness of metals and inhibitor molecule is given by ηm and ηi, respectively. In case of iron the theoretical values of χ_m and ηm are 7 and 0 eV mol^{-1} respectively [57]. However, the value of 7.0 eV corresponds to the free electron gas Fermi energy of iron in the free electron gas model. But the use of this value for χ_m is conceptually wrong because here electron-electron interaction has been neglected [58]. Thus, work function (φ_m) of the metal has been incorporated for an appropriate measurement of electronegativity. So, the new formula for the estimation of N is as follows [59]:

$$\Delta N = \frac{\varphi_m - \chi_i}{2(\eta_m - \eta_i)} \tag{13}$$

The work function (φ) values for iron surface plan Fe (1 0 0), Fe (1 1 0) and Fe (1 1 1) surfaces are 3.91, 4.82 and 3.88 eV, respectively [59, 60]. The surface energy of iron planes for bcc structure are in order of Fe (110) < Fe (100) < Fe (111) and out of these the most stable on plan is Fe (110). Thus, most commonly Fe (110) plan is used in corrosion inhibition research [61].

4.7 Fukui function and local softness

In the field of corrosion research inhibition of corrosion is governed through donation and acceptance of electron, which basically involves the nucleophilic and electrophilic reactions. This can be achieved through the evaluation of the Fukui indices [61–63]. Thus, by invoking the HSAB principle in a local sense, one may establish the behavior of the different sites with respect to hard or soft reagents.

The Fukui function generated using the finite difference approximation is as follows:

$$f^+ = \left(\frac{\delta\rho(r)}{\delta N}\right)_\nu^+ = \rho_{(N+1)} - \rho_N \tag{14}$$

$$f^- = \left(\frac{\delta\rho(r)}{\delta N}\right)_\nu^+ = \rho_{(N)} - \rho_{(N-1)} \tag{15}$$

where f^+ is nucleophilic and f^- is electrophilic Fukui functions respectively, $\rho_{(N+1)}$, $\rho_{(N)}$ and $\rho_{(N-1)}$ are the electronic densities of anionic, neutral and cationic forms of the atoms with *N+1*, *N* and *N-1* electrons.

Castro et al. [64] studied the series of isomers of the Schiff base (E)-2-(2-hydroxybenzylideneamino) phenylarsonic acid as corrosion inhibitor. The author has pictorially presented the negative Fukui, i.e., f^- function and is given as follows.

In **Figure 4**, the black region represents the area where inhibitor contains the major portion of electrons for donating to the empty d orbitals of Fe metal. The Fukui function represents the change of the electronic density in a given point with

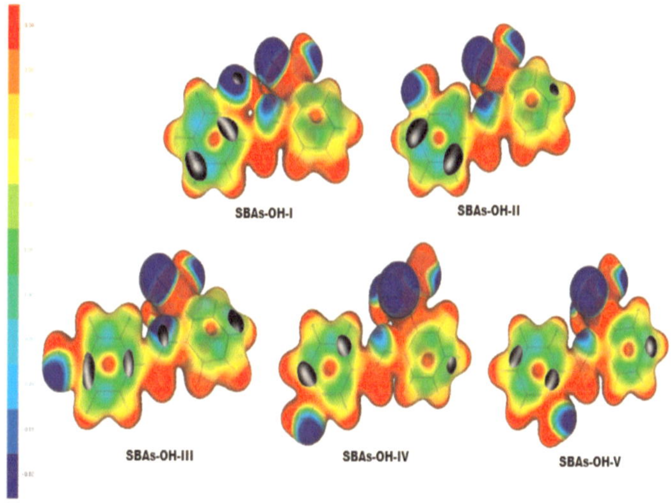

Figure 4.
Isosurface plots for the electrophilic form of the Fukui function f^-.

respect to the change in the number of electrons N [65]. For the electrophilic attack to occur, the contributing atoms are likely to be the carbon atoms on the OH-substituted phenyl ring in all isomers, the oxygen atom in the isomer SBAs-OH-I, the nitrogen atom of the imine group of isomer SBAs-OH-III, and a region located in AsO(OH)2-substituted phenyl ring.

Local softness is calculated is using the equation [66]:

$$s(r) = \left[\frac{\partial n(r)}{\partial \mu}\right]_{\nu(r)} \tag{16}$$

The integration of equation gives global softness value:

$$S = \int s(r)dr \tag{17}$$

Local softness and Fukui function are related as per below equation:

$$s(r) = Sf(r) \tag{18}$$

The local softness contains information similar to those obtained from Fukui function plus additional in-formation about the molecular softness, which is related to the global reactivity with respect to a reaction partner.

4.8 Charge transfer: donation and back donation

The energy change in the molecule is attributed because of two processes either electron donation to the metal or acceptance from the metal, i.e., back donation. Gomez et al. [67] proposed a simple charge transfer model for donation and back donation. In this model when the molecule receives certain amount of charge (ΔN^-) the change in energy is given as follows:

$$\Delta E^+ = \delta^+ \Delta N^+ + 0.5\eta(\Delta N^+)^2 \tag{19}$$

However, when the molecule back donate certain amount of charge (ΔN^-) the change in energy is given as follows:

$$\Delta E^{-} = \delta^{-}\Delta N^{-} + 0.5\eta(\Delta N^{-})^{2} \quad (20)$$

If the total change in energy (ΔE_T) is the summation of ΔE^+ and ΔE^-, and assuming that the donation and back donation, i.e., (ΔN^-) and (ΔN^+) are equal, than:

$$\Delta E_T = \Delta E^{-} + \Delta E^{+} = (\delta^{-} - \delta^{-})\Delta N^{+} + \eta(\Delta N^{+})^{2} \quad (21)$$

In a situation when the total energy change becomes a minimum with respect to ΔN^+, then

$$\Delta N^{+} = -(\delta^{+} - \delta^{-})/2\eta \quad (22)$$

and change in total energy becomes as follows:

$$\Delta E_T = -(\delta^{+} - \delta^{-})/4\eta = -\eta/4 \quad (23)$$

According to Guo et al. [68] the change in total energy are negative, i.e., $\Delta E_T < 0$. This implies that the back donation of charge from metal surface to inhibitor molecules is favorable.

5. Inhibitor/surface interaction study using molecular dynamic simulation (MD)

Molecular dynamics is a computer-based modeling technique by which the evolution as a function of time or trajectory of a molecule is described by the principles of classical Newtonian mechanics [69]. In molecular dynamics, intramolecular movements of inhibitor molecules are simulated in order to visualize the real picture of corrosion inhibition process. Molecular dynamics simulation calculation has two serve two purposes:

a. Intramolecular movements' simulation and calculation of thermodynamic properties such as entropy, free energy, etc.

b. Optimization of the molecular structures efficiently by avoiding multiple minima by annealing the simulation process. The range of simulation time is approximately between 10^{-14} and 10^{-10}s per simulation.

Molecular dynamics simulation involves the implementation of initial steps: In the first very step build molecular structure are introduced or immersed in a box having solvent molecules, after that simulation process starts. The total energy of the molecular system should be minimizing before the start of the simulation, in order to avoid the generation of aberrant trajectories (when the initial forces are too great). For achieving this, the steepest slope/steepest descent method and that of the conjugated gradients have been frequently used.

6. Corrosion inhibitors studied using DFT

6.1 Adsorption behavior of inhibitors on mild steel surface

In literature, several DFT calculations are observed where inhibition efficiency and molecular structure/electronic properties of organic corrosion inhibitors have been correlated. Xu et al. [70] has studied corrosion inhibition property of Schiff's base derivatives via DFT calculation using GGA/BLYP method. The result indicates that 2-PCT molecules adsorb strongly onto iron surface and provide higher inhibition efficiency.

Zhang et al. [71] investigated the inhibition of methionine (Met) and proline (Pro) in PCMs solution using DFT and B3LYP/6-31G (d) method. The inhibition efficiency of Met and Pro follow the order of Met > Pro.

The corrosion inhibition effects of thiourea (TU), methylthiourea (MTU) and phenylthiourea (PTU) on mild steel in 0.1MH2SO4 was investigated by Ozcan et al. [72]. The electronic properties were calculated using ab initio RHF/6-31G(d) method. The energy of E_{HOMO}, the energy of E_{LUMO} and energy gap values supports that PTU is best inhibitor.

The corrosion inhibition performance of three naphthyridines derivatives were explored for mild steel in 1 M HCl by Singh et al. [73] using DFT methods. The inhibition efficiencies of N-3 and N-2 are higher due to the presence of electron-donating $-OCH_3$ and $-CH_3$ substituents respectively as compared to N-1, which has no substituents.

Feng et al. [74], studied inhibition activity of an imidazoline derivative, namely 1-[N,N′-bis(hydroxylethylether)- aminoethyl]-2-stearicimidazoline (HASI) for carbon steel in 5% NaCl saturated with $Ca(OH)_2$ solution using DMol3 module. DFT calculations reveals that N=C-N region in imidazoline ring is the most active reaction site.

6.2 Adsorption behavior of inhibitors on aluminum surface

The inhibitive action of poly ethylene glycol (PEG) against the corrosion of aluminum surface in 1 M HCl was investigated by Awad et al. [75] using B3LYP/6–31 + G(d,p) basis set. DFT was performed on repeating units of 1, 2, 3, 4 and 5. The inhibition performance increases with increase in number of repeating unit.

The author Li et al. [76] studied the corrosion inhibition effect of three oxime compounds on the corrosion of aluminum in HCl solution using BLYP in conjunction with double numerical plus d-functions (DND) basis set. In these inhibitors, HOMO and LUMO electron densities are localized principally on C=N–OH, which suggest that C=N–OH functional group could both accept and donate the electron.

Khaled et al. [77], investigated the corrosion effect of and adsorption characteristics of three imidazole derivatives on aluminum in 1.0 M HCl using local density functional (LDF) method with a double numeric polarization (DNP) basis set, and a Becke–Perdew (BP) functional. A good correlation between the rate of corrosion and EHOMO, as well as with energy gap ($\Delta E = E_{\mathrm{LUMO}} - E_{\mathrm{HOMO}}$).

Kaya et al. [78] applied HF and DFT/B3LYP methods with SDD, 6-31++G (d, p) and 6-31 G basis sets both in gaseous and aqueous phase for studying the adsorption and corrosion inhibition properties of some benzotriazole and phospono derivatives on aluminum. The inhibition efficiency ranking of these molecules as: PBA > PBTA>PAA > TBTA.

6.3 Adsorption behavior of inhibitors on copper surface

Corrosion inhibition of indazole (IA) and 5-aminoindazole (AIA) for copper in NaCl solution was studied by Qiang et al. [79]. The molecular structures of IA and AIA were geometrically optimized by density functional theory (DFT) using B3LYP functional with 6-311++G(d,p) basis set in aqueous phase. The higher value of HOMO and lower value of ΔE in case of AIA concludes that the interaction of AIA with copper will be stronger than that between IA and copper.

Khalid and Madkour et al. [13, 80, 81] probe the interactions between benzotriazole, methionine, phenol derivatives and copper via quantum calculations respectively. The obtained results suggests that quantum parameters are very useful in characterizing organic compounds as an adsorbate.

7. Corrosion inhibitors studied using MD

Feng et al. [74] performed the MD simulation of 1-[N,N′-bis (hydroxylethylether)- aminoethyl]-2-stearicimidazoline (HASI). The MD calculation suggests that adsorbed imidazoline molecule is parallel to the iron surface in order to maximize its contact with the surface (**Figure 5**). The interaction energy for Fe atom is −284 kJ/mol, for Fe_3O_4 is −226 kJ/mol and for Fe_2O_3 is −157 kJ/mol. So, inhibitor adsorbed on Fe surface more strongly than Fe_3O_4 and Fe_2O_3.

MD calculation of adsorption of indazole (IA) and 5-aminoindazole (AIA) on copper surface was studied Qiang et al. [79]. The optimized equilibrium configuration of inhibitors molecules are shown in **Figure 6** and inhibitors adsorption occurs through parallel orientation. The adsorption energy between Cu (1 1 1) surface and inhibitor molecule are −250.86 kJ/mol for IA and −307.86 kJ/mol for AIA.

Molecular level MD calculation of Benzotriazole and Phospono derivatives was studied by Kaya et al. [78] on aluminum surface. The best adsorption configuration of inhibitor molecules on Al (111) surface are represented in **Figure 7**. The parallel configuration of inhibitor molecules suggests the stronger adsorption.

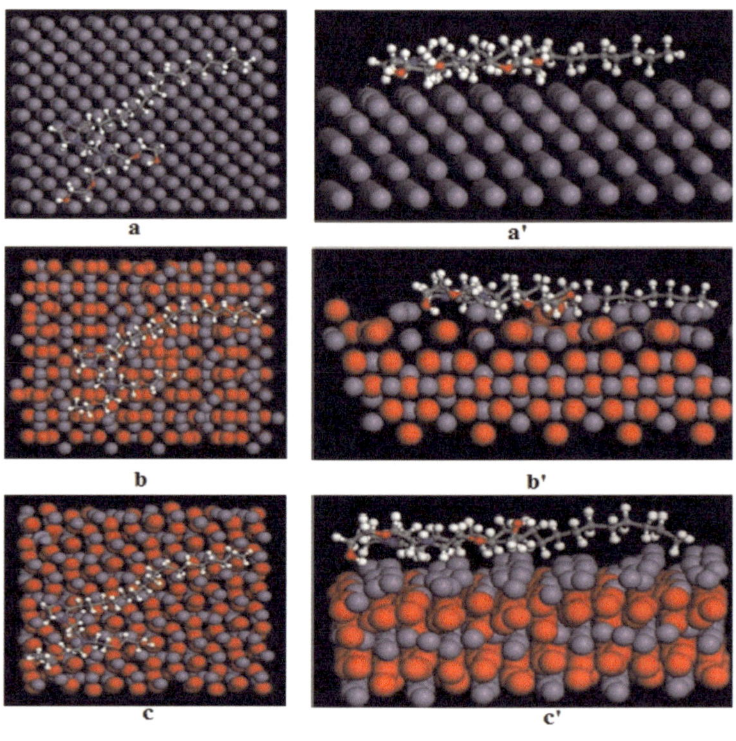

Figure 5.
Adsorption of imidazoline molecule on (a) Fe, (b) Fe_3O_4 and (c) Fe_2O_3 surfaces. The left is top view, and the right is side view [74].

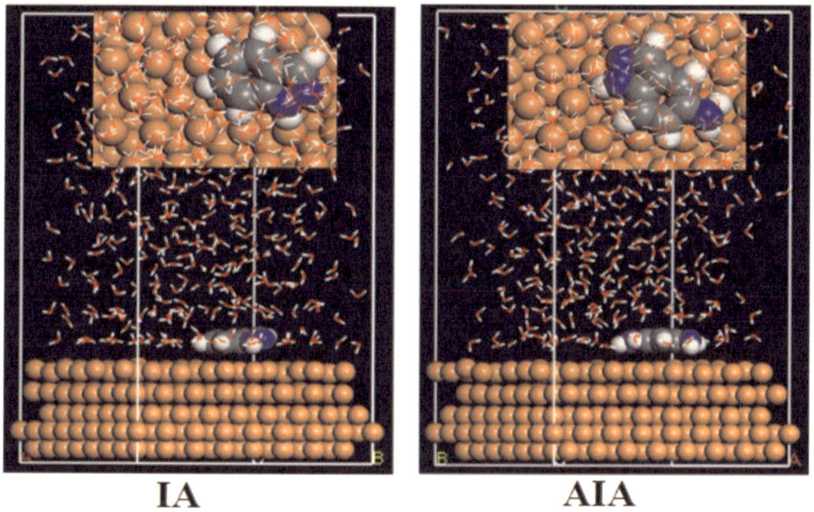

Figure 6.
Equilibrium adsorption configuration for IA and AIA on Cu (111) surface in 3% NaCl solution after optimization (inset images show the on-top views) [79].

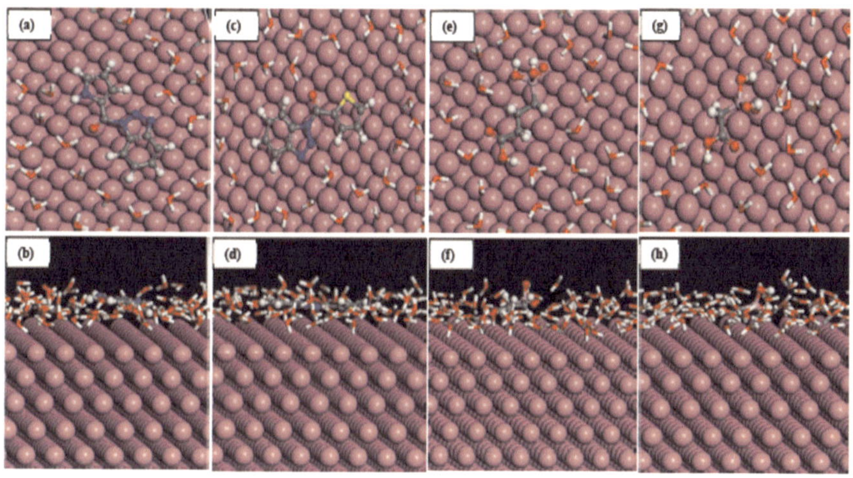

Figure 7.
Equilibrium adsorption configuration of inhibitors PBTA (a and b), TBTA (c and d), BPA (e and f), PAA (g and h) on Al (1 1 1) surface. Top: top view and bottom: side view [78].

8. Conclusion and future work

Computational program provide an atomistic view for understanding the corrosion inhibition process and significantly contribute for the selection of adequate inhibitor molecules. DFT calculations has made possible to characterize a suitable inhibitor molecules using their intrinsic properties such as energies of HOMO and LUMO orbitals, energy gap (ΔE), electronegativity, hardness, softness, etc. It helps corrosion scientist to understand the inhibitor-metal interaction through the locations of HOMO and LUMO orbitals. The examination of local reactivity reveals the nucleophilic and electrophilic centers which react with the metal surfaces.

The literature survey suggests that with help of quantum chemistry a good understanding of corrosion system can be achieved when various metal surfaces and various inhibitor molecules are compared. DFT has ability to perform calculations for complex or large organic molecules that can be applied as a potential corrosion inhibitor. In addition, DFT provide a platform to scientist for designing a unique inhibitor molecule and understand their molecular structure at atomistic level.

In recent years, computational modeling like DFT and MD provide real inhibitor-metal interaction picture by performing calculation in aqueous phase. Such an approach act as a benchmark for experimental study in the field of corrosion research.

Molecular dynamic simulation (MD) gives a representative image of inhibitor molecule orientation on the metal surface. This helps to analyze the inhibition activity of inhibitor molecules, i.e., inhibitor with parallel or flat orientation with respect to metal surface will cover more surface area and provide a better corrosion inhibition protection as compared to the inhibitor which have vertical orientation.

Future research includes modeling the electrochemical potential, generation of alloy surface layers, addition of thin oxide film covering onto the metal surface, calculation of more complex molecule, comprehension of solvent molecule role in adsorption, understanding the defect role in adsorption process, etc.

Acknowledgements

The authors are thankful to the Sichuan 1000 Talent Fund and financial assistance provided by the National Natural Science Foundation of China (No. 51274170).

Conflict of interest

There are no conflicts to declare.

Author details

Ambrish Singh[1,2*], Kashif R. Ansari[3], Mumtaz A. Quraishi[3] and Yuanhua Lin[1,2]

1 School of Materials Science and Engineering, Southwest Petroleum University, Chengdu, Sichuan, China

2 State Key Laboratory of Oil and Gas Reservoir Geology and Exploitation, Southwest Petroleum University, Chengdu, Sichuan, China

3 Centre of Research Excellence in Corrosion, Research Institute, King Fahd University of Petroleum and Minerals, Dhahran, Saudi Arabia

*Address all correspondence to: vishisingh4uall@gmail.com

References

[1] Mao ZJ, Jun L. Corrosion inhibition performance of carbon steel in brine solution containing H_2S and CO_2 by novel Gemini surfactants. Acta Physico-Chimica Sinica. 2012;**28**:623-629

[2] Obot IB, Ebenso EE, Kabanda MM. Metronidazole as environmentally safe corrosion inhibitor for mild steel in 0.5 M HCl: Experimental and theoretical investigation. Journal of Environmental Chemical Engineering. 2013;**1**:431-439

[3] Trabanelli G. Chemical Industries Corrosion Mechanism. Vol. 3. New York: Marcel Dekker; 1987

[4] Gutiérreza E, Rodrígueza JA, Borbollaa JC, Rodrígueza JGA, Thangarasu P. Development of a predictive model for corrosion inhibition of carbonsteel by imidazole and benzimidazole derivatives. Corrosion Science. 2016;**108**:23-35

[5] Shaban SM. N-(3-(dimethyl benzyl ammonio)propyl) alkanamide chloride derivatives as corrosion inhibitors for mild steel in 1 M HCl solution: Experimental and theoretical investigation. RSC Advances. 2016;**6**: 39784-39800

[6] Ramaganthan B, Gopiraman M, Olasunkanmi LO, Kabanda MM, Yesudass S, Bahadur I, et al. Synthesized photo-cross-linking chalcones as novel corrosion inhibitors for mild steel in acidic medium: Experimental, quantum chemical and Monte Carlo simulation studies. RSC Advances. 2015;**5**: 76675-76688

[7] Obot IB, Macdonald DD, Gasem ZM. Density functional theory (DFT) as a powerful tool for designing new organic corrosion inhibitors. Part 1: An overview. Corrosion Science. 2015;**99**:1-30

[8] Verma C, Lgaz H, Verma DK, Ebenso EE, Bahadur I, Quraishi MA. Molecular dynamics and Monte Carlo simulations as powerful tools for study of interfacial adsorption behavior of corrosion inhibitors in aqueous phase: A review. Journal of Molecular Liquids. 2018;**260**: 99-120

[9] Yang W, Parr RG, Pucci R. Electron density, Kohn–Sham frontier orbitals, and Fukui functions. The Journal of Chemical Physics. 1954;**81**:2862-2863

[10] Parr RG, Yang W. Density functional approach to the frontier-electron theory of chemical reactivity. Journal of the American Chemical Society. 1984;**106**:4049-4050

[11] Armaković S, Armaković SJ, Vraneš M, Tot A, Gadžurić S. DFT study of 1-butyl-3-methylimidazolium salicylate: A third-generation ionic liquid. Journal of Molecular Modeling. 2015;**21**:246-256

[12] Guo L, Zhang ST, Lv TM, Feng WJ. Comparative theoretical study on the corrosion inhibition properties of benzoxazole and benzothiazole. Research on Chemical Intermediates. 2015;**41**:3729-3742

[13] Madkour LH, Elshamy IH. Experimental and computational studies on the inhibition performances of benzimidazole and its derivatives for the corrosion of copper in nitric acid. International Journal of Industrial Chemistry. 2016;**7**:195-221

[14] Mousavi M, Mohammadalizadeh M, Khosravan A. Theoretical investigation of corrosion inhibition effect of imidazole and its derivatives on mild steel using cluster model. Corrosion Science. 2011;**53**:3086-3091

[15] Camacho RL, Montiel E, Jayanthi N, Pandiyan T, Cruz J. DFT studies of alpha-diimines adsorption over Fe-n surface (n = 1, 4, 9 and 14) as a model

for metal surface coating. Chemical Physics Letters. 2010;**485**(1–3):142-151

[16] Wang Y, Cheng X, Yang X, Yang X. DFT study of solvent effects for some organic molecules using a polarizable continuum model. Journal of Solution Chemistry. 2006;**35**:869-878

[17] Dogonadze RR. Theory of molecular electrode kinetics. In: Hush NS, editor. Reactions of Molecules at Electrodes. London: Interscience Pub.; 1971. pp. 135-227

[18] Sholl D, Steckel JA. Density Functional Theory: A Practical Introduction. New York: Wiley; 2009

[19] Enge E l, Dreizler RM. Density Functional Theory: An Advanced Course. Berlin, Heidelberg: Springer-Verlag; 2011

[20] Parr RG, Weitao Y. Density-Functional Theory of Atoms and Molecules. USA: Oxford University Press; 1989

[21] Marques MAL, Maitra NT, Nogueira FMS, Gross EKU, Rubio A. Fundamentals of Time-Dependent Density Functional Theory. Berlin, Heidelberg: Springer-Verlag; 2012

[22] Fiolhais C, Marques MAL, Nogueira F. A Primer in Density Functional Theory. Berlin, Heidelberg: Springer-Verlag; 2003

[23] Sahni V. Quantal Density Functional Theory. Berlin, Heidelberg: Springer Science & Business Media; 2004

[24] Tsuneda T. Density Functional Theory in Quantum Chemistry. Japan: Springer; 2014

[25] Gross E, Dreizler RM. Density Functional Theory: An Approach to the Quantum Many-Body Problem. Berlin, Heidelberg: Springer; 1990

[26] Gross EKU, Dreizler RM. Density Functional Theory. New York: Springer Science & Business Media; 1995

[27] Amador C. Density Functional Theory. Berlin, Heidelberg: Springer-Verlag; 1983

[28] Hohenberg P, Kohn W. Inhomogeneous electron gas. Physics Review B. 1964;**136**:864-871

[29] Koch W, Holthausen MC. A Chemist's Guide to Density Functional Theory. Weinheim: Wiley-VCH; 2000

[30] Bel SL, Dines TJ, Chowdhry BZ, Withnall R. Computational chemistry using modern electronic structure methods. Journal of Chemical Education. 2007;**84**:1364-1368

[31] Kohn W, Sham L. Self-consistent equations including exchange and correlation effects. Physics Review. 1965;**140**:A1133-A1138

[32] Perdew JP. Density functional approximation for the correlation energy of an inhomogeneous electron gas. Physical Review B. 1986;**33**: 8822-8824

[33] Becke AD. Density functional calculations of molecular bond energies. The Journal of Chemical Physics. 1986; **84**:4524-4529

[34] Lee C, Yang W, Parr RG. Development of the Colle-Salvetti correlation energy formula into a functional of the electron density. Physical Review B. 1988;**37**:785-789

[35] Perdew J, Yang W. Accurate and simple density functional for the electronic exchange energy: Generalized gradient approximation. Physical Review B. 1986;**33**:8800-8802

[36] Becke AD. Density–functional exchange-energy approximation with

correct asymptotic behavior. Physical Review A. 1988;**38**:3098-3100

[37] Ramachandran KI, Deepa G, Namboori K. Computational chemistry and molecular modeling, Principles and Applications. Berlin, Heidelberg: Springer; 2008;137

[38] Arshad MN, Bibi A, Mahmood T, Asiri AM, Ayub K. Crystal structures and spectroscopic properties of triazine based hydrazone derivatives; a comparative experimental-theoretical study. Molecules. 2015;**20**:5851-5874. DOI: 10.3390/molecules20045851

[39] Zhan CG, Nichols JA, Dixon DA. Ionization potential, electron affinity, electronegativity, hardness, and electron excitation energy: Molecular properties from density functional theory orbital energies. The Journal of Physical Chemistry. A. 2003;**107**: 4184-4195

[40] Singh A, Ansari KR, Kumar A, Liu W, Songsong C, Lin Y. Electrochemical, surface and quantum chemical studies of novel imidazole derivatives as corrosion inhibitors for J55 steel in sweet corrosive environment. Journal of Alloys and Compounds. 2017;**712**: 121-133

[41] Tian H, Li W, Cao K, Hou B. Potent inhibition of copper corrosion in neutral chloride media by novel non-toxic thiadiazole derivatives. Corrosion Science. 2013;**73**:281-291

[42] Hegazy MA, Badawi AM, Abd El Rehim SS, Kamel WM. Corrosion inhibition of carbon steel using novel N-(2-(2-mercaptoacetoxy)ethyl)-N,N-dimethyl dodecan-1-aminium bromide during acid pickling. Corrosion Science. 2013;**69**:110-122

[43] Kovacevic N, Kokalj A. The relation between adsorption bonding and corrosion inhibition of azole molecules on copper. Corrosion Science. 2013;**73**: 7-17

[44] Deng S, Li X, Xie X. Hydroxymethyl urea and 1,3-bis(hydroxymethyl) urea as corrosion inhibitors for steel in HCl solution. Corrosion Science. 2014;**80**: 276-289

[45] Singh A, Ansari KR, Haque J, Dohare P, Lgaz H, Salghi R, et al. Effect of electron donating functional groups on corrosion inhibition of mild steel in hydrochloric acid: Experimental and quantum chemical study. Journal of the Taiwan Institute of Chemical Engineers. 2018;**82**:233-251

[46] Mendoza-Huizar LH, Rios-Reyes CH. Chemical reactivity of atrazine employing the Fukui function. Journal of the Mexican Chemical Society. 2011; **55**:142-147

[47] Chermette H. Chemical reactivity indexes in density functional theory. Journal of Computational Chemistry. 1999;**20**:129-154

[48] Pearson RG. Absolute electronegativity and hardness correlated with molecular orbital theory. Proceedings of the National Academy of Sciences. 1986;**83**: 8440-8441

[49] Parr RG, Gadre SR, Bartolotti LJ. Local density functional theory of atoms and molecules. Proceedings of the National Academy of Sciences of the United States of America. 1979;**76**: 2522-2526

[50] Haque J, Ansari KR, Srivastava V, Quraishi MA, Obot IB. Pyrimidine derivatives as novel acidizing corrosion inhibitors for N80 steel useful for petroleum industry: A combined experimental and theoretical approach. Journal of Industrial and Engineering Chemistry. 2017;**49**:176-188

[51] Zevallos J, Labbé AT. A theoretical analysis of the Kohn-Sham and Hartree-Fock orbitals and their use in the determination of electronic properties.

Journal of the Chilean Chemical Society. 2003;**48**:39-47

[52] Ghanty TK, Ghosh SK. Correlation between hardness, polarizability, and size of atoms, molecules, and clusters. The Journal of Physical Chemistry. 1993; **97**:4951-4953

[53] Chattaraj PK, Sengupta S. Popular electronic structure principles in a dynamical context. The Journal of Physical Chemistry. 1996;**100**: 16126-16130

[54] Hohm U. Is there a minimum polarizability principle in chemical reactions? The Journal of Physical Chemistry. A. 2000;**104**:8418-8423

[55] Parr RG, Sventpaly L, Liu S. Electrophilicity index. Journal of the American Chemical Society. 1999;**121**: 1922-1924

[56] Geerlings P, De Proft F, Langenaeker W. Conceptual density functional theory. Chemical Reviews. 2003;**103**:1793-1873

[57] Pearson RG. Absolute electronegativity and hardness: Application to inorganic chemistry. Inorganic Chemistry. 1988;**27**:734-740

[58] Kovacˇević N, Kokalj A. DFT study of interaction of azoles with Cu(111) and Al(111) surfaces: Role of azole nitrogen atoms and dipole-dipole interactions. Journal of Physical Chemistry C. 2011;**115**:24189-24197

[59] Kokalj A. On the HSAB based estimate of charge transfer between adsorbates and metal surfaces. Chemical Physics. 2012;**393**:1-12

[60] Cao Z, Tang Y, Cang H, Xu J, Lu G, Jing W. Novel benzimidazole derivatives as corrosion inhibitors of mild steel in the acidic media. Part II: Theoretical studies. Corrosion Science. 2014;**83**:292-298

[61] Kokalj A. Is the analysis of molecular electronic structure of corrosion inhibitors sufficient to predict the trend of their inhibition performance. Electrochimica Acta. 2010;**56**:745-755

[62] Lewars EG. Computational Chemistry Introduction to the Theory and Applications of Molecular and Quantum Mechanics. London: Springer; 2011

[63] Oláh J, Alsenoy CV. Condensed Fukui functions derived from stockholder charges: Assessment of their performance as local reactivity descriptors. The Journal of Physical Chemistry. A. 2002;**106**:3885-3890

[64] Castro ME, Percino MJ, Cerón M, Soriano G, Chapela VM. Theoretical inhibition efficiency study of Schiff Base (E)-2-(2-hydroxybenzylideneamino) phenylarsonic acid and its isomers. International Journal of Electrochemical Science. 2014;**9**:7890-7903

[65] López P, Méndez F. Fukui function as a descriptor of the imidazolium protonated cation resonance hybrid structure. Organic Letters. 2004;**6**: 1781-1783

[66] Yang W, Parr RG. Hardness, softness and the Fukui function in the electronic theory of metals and catalysis. Proceedings of the National Academy of Sciences of the United States of America. 1985;**82**:6723-6726

[67] Gomez B, Likhanova NV, Dominguez-Aguilar MA, Martinez-Palou R, Vela A, Gazquez JL. Quantum chemical study of the inhibitive properties of 2-pyridyl-azoles. The Journal of Physical Chemistry. B. 2006; **110**:8928-8934

[68] Guo L, Ren X, Zhou Y, Xu S, Gong Y, Zhang S. Theoretical evaluation of the corrosion inhibition performance of 1,3-thiazole and its amino derivatives.

Arabian Journal of Chemistry. 2017;**10**: 121-130

[69] Rives JT, Jorgensen WL. Molecular dynamics of proteins with the OPLS potential functions. Simulation of the third domain of silver pheasant ovomucoid in water. Journal of the American Chemical Society. 1990;**112**: 2773-2781

[70] Xu B, Yang W, Liu Y, Yin X, Gong W, Chen Y. Experimental and theoretical evaluation of two pyridinecarboxaldehyde thiosemicarbazone compounds as corrosion inhibitors for mild steel in hydrochloric acid solution. Corrosion Science. 2014;**78**:260-268

[71] Zhang Z, Tian N, Zhang W, Huang X, Ruan L, Wu L. Inhibition of carbon steel corrosion in phase-change-materials solution by methionine and proline. Corrosion Science. 2016;**111**: 675-689

[72] Ozcan M, Dehri I, Erbil M. Organic sulphur-containing compounds as corrosion inhibitors for mild steel in acidic media: Correlation between inhibition efficiency and chemical structure. Applied Surface Science. 2004;**236**:155-164

[73] Singh P, Ebenso EE, Olasunkanmi LO, Obot I, Quraishi M. Electrochemical, theoretical, and surface morphological studies of corrosion inhibition effect of green naphthyridine derivatives on mild steel in hydrochloric acid. Journal of Physical Chemistry C. 2016;**120**:3408-3419

[74] Feng L, Yang H, Wang F. Experimental and theoretical studies for corrosion inhibition of carbon steel by imidazoline derivative in 5% NaCl saturated $Ca(OH)_2$ solution. Electrochimica Acta. 2011;**58**:427-436

[75] Awad M, Metwally M, Soliman S, El-Zomrawy A. Experimental and quantum chemical studies of the effect of poly ethylene glycol as corrosion inhibitors of aluminum surface. Journal of Industrial and Engineering Chemistry. 2014;**20**:796-808

[76] Li X, Deng S, Xie X. Experimental and theoretical study on corrosion inhibition of oxime compounds for aluminium in HCl solution. Corrosion Science. 2014;**81**:162-175

[77] Khaled K, Amin MA. Electrochemical and molecular dynamics simulation studies on the corrosion inhibition of aluminum in molar hydrochloric acid using some imidazole derivatives. Journal of Applied Electrochemistry. 2009;**39**: 2553-2568

[78] Kaya S, Banerjee P, Saha SK, Tüzün B, Kaya C. Theoretical evaluation of some benzotriazole and phospono derivatives as aluminum corrosion inhibitors: DFT and molecular dynamics simulation approaches. RSC Advances. 2016;**6**:74550-74559

[79] Qiang Y, Zhang S, Xu S, Li W. Experimental and theoretical studies on the corrosion inhibition of copper by two indazole derivatives in 3.0% NaCl solution. Journal of Colloid and Interface Science. 2016;**472**:52-59

[80] Khaled K, Amin MA, Al-Mobarak N. On the corrosion inhibition and adsorption behaviour of some benzotriazole derivatives during copper corrosion in nitric acid solutions: A combined experimental and theoretical study. Journal of Applied Electrochemistry. 2010;**40**:601-613

[81] Khaled K. Corrosion control of copper in nitric acid solutions using some amino acids—A combined experimental and theoretical study. Corrosion Science. 2010;**52**:3225-3234

Section 3

Tribocorrosion

Chapter 8

Electrochemical Techniques for Corrosion and Tribocorrosion Monitoring: Fundamentals of Electrolytic Corrosion

Abdenacer Berradja

Abstract

This first chapter aims at giving a brief framework for understanding the principles of electrochemistry in corrosion and tribocorrosion, enabling potential users and readers to quickly apprehend the electrochemical nature of corrosion. The subsequent chapter will provide additional details on the methods derived from these principles in a clear manner and ready-to-use format. In particular, the implementation of electrochemical techniques for the study of tribocorrosion allows, *in situ* and in real time, to monitor and control the corrosion conditions during wear and to quantify the corrosion kinetics. An introductory to some of the basic terms and concepts of electrochemistry and corrosion is first presented. Then, an overview of the thermodynamic and kinetic parameters of relevance to corrosion electrochemistry is highlighted. A description on how the electrical nature of corrosion reactions allows the interface to be modeled as an electrical circuit, as well as how this electrical circuit can be used to obtain information on corrosion rates. These prerequisites are necessary to better understand the surface reactivity of metals and other electronic conductive materials immersed in ionic electrolyte media whether or not subjected to mechanical stimuli.

Keywords: corrosion, tribo-electrochemistry, tribocorrosion, standard potential, electrode kinetics

1. Introduction

The corrosion science is a complex subject that is not well defined, and still continues to progress as the subject evolves from the simple traditional definition of "destruction of metals through oxidation and its prevention" to "degradation of a material that involves one or more chemical and/or electrochemical reactions and its foresight". This latter material definition encompasses a wide range of environments, and all classes of materials (ceramics, organics, composites), not just metals. The intentional conjunction with degradation due to non-chemical processes, such as tribology (i.e., science of friction, wear and lubrication), has open up a new global perspective topic of a rare universality that leads us to prospect many aspects of science and technology, namely tribocorrosion. To better understand the specificity of tribocorrosion, it is necessary to briefly recall what distinguishes corrosion from tribocorrosion.

Corrosion deals primarily with the electrochemical aspects related to physical-chemical oxidation and reduction processes taking place at the surface of materials and the effect of the reactivity of surfaces of materials with respect to their environment, time, pH, temperature, pressure, and electrolyte composition. It is almost without exception an irreversible heterogeneous reaction of a material with the environment, which usually (but not always) results in a degradation of the material or its properties (e.g., decadence of the functional properties of materials). Examples of corrosion phenomena include the transformation of steel into rust, oxidation of an electrical copper contact, cracking of brass in the presence of ammonia, pipeline degradation by H_2S, swelling of PVC in contact with a solvent, alkaline attack on refractory bricks, and mineral glasses. Occasionally, in certain cases, corrosion is valuable. For instance, the disposal of neglected metallic objects in the nature is not an uncommon corrosion phenomenon. It is still common to find beneficial corrosion reactions in the field practice. A typical example of corrosion protection processes is the anodizing of aluminum. Anodizing strengthens the passive oxide film on the surface of aluminum, and therefore its resistance to corrosion, but also serves as a decorative effect. Likewise, corrosion reactions are used to produce a smooth surface finish in chemical and electrochemical polishing processes.

Tribocorrosion occurs when surfaces of materials subjected to mechanical contacts and in relative motion and/or behaviors are affected by chemical, electrochemical and/or biological environmental factors. Accordingly, tribocorrosion damage (i.e., material loss) can be designated in a broad sense as a failure mechanism due to the mutual interaction of corrosion, friction, and wear processes and their synergy effects. It generates changes in surface and/or volume compositions (e.g., alteration of materials properties, surface and sub-surface transformations, cracking, tribo-chemical reactions, etc.), and often modifies the environment (e.g., surface contamination by tribo-reaction products or corrosion-produced compounds, pH changes, and so on), and ultimately can lead to system failure. Such a physical-chemical deterioration is well described in the literature by the following terms: corrosive wear, fretting-corrosion or corrosion-erosion [1–6]. Corrosion and wear often combine to cause aggressive damage and at last a shutdown in a number of industries, such as mining, mineral processing, chemical processing, metal components of machines, marine structures, pulp and paper production, ships, bridges, biomechanics (e.g., orthopedics), civil engineering structures and energy production, and so on. Although, corrosion can often occur in the absence of mechanical wear, the opposite is rarely true. Corrosion accompanies to some extent in all environments, except in vacuum and inert atmospheres. The combined effects of friction (wear) and corrosion can result in total material losses well above than that of the additive effects of each process taken apart, which is attributed to their synergy [1–3, 7]. These effects are still difficult to control. Knowledge of the tribological behavior of a material couple in the absence of any chemical attack and the knowledge of the electrochemical behavior in the absence of any mechanical impact are not sufficient to deduce the tribocorrosion behavior of that couple system of materials. In many articulation systems, it has been noticed that friction may alter the sensitivity, and modifies the composition of the surfaces of materials in moving contacts to corrosion. In turn, corrosion can affect the friction (and wear) process of moving contacting parts. This usually accelerates the tribochemical degradation of the material, which may affect the contact moving conditions, and thus the friction process and the coefficient of friction too [1–6]. The contact motion can be a continuous or discontinuous one; it can be a unidirectional or a reciprocating one. The complexity of a tribocorrosion system is illustrated in **Figure 1**.

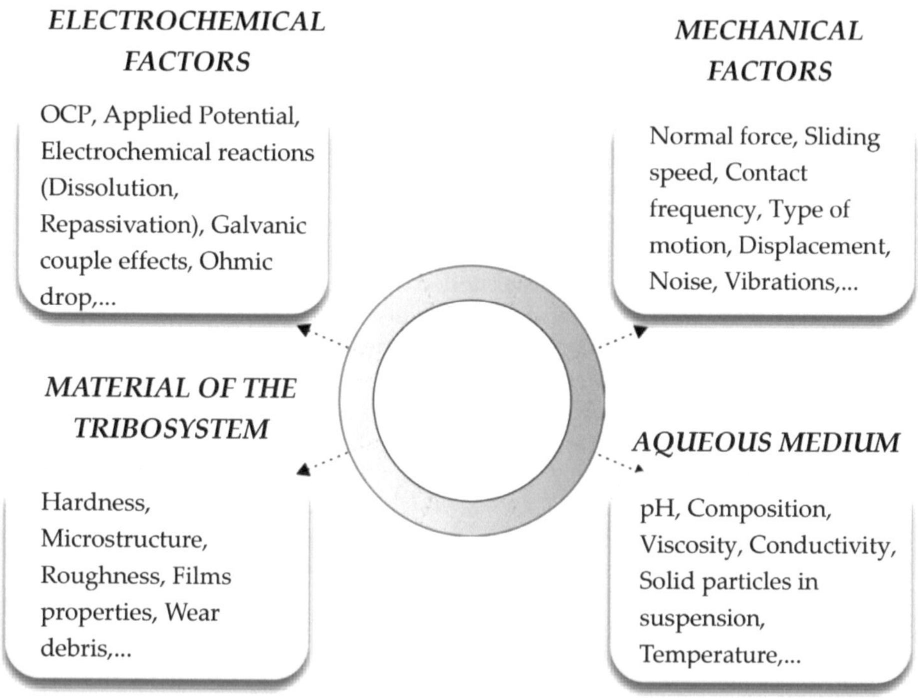

Figure 1.
Schematic of main factors which define the complexity of a tribocorrosion system.

Tribocorrosion of moving solids immersed in an aggressive environment is, just as friction, a system parameter. The tribo-electrochemical system consists of implementing *in-situ* electrochemical techniques to a tribological designed system under well-controlled conditions (i.e., tribometer type, and complete material system) [3]. A properly designed laboratory mechanical system allows for the best simulation as much as possible the entire material system used in the field, and the constraints that have associated with it (e.g., vibration mode, similarity of the wear mechanisms active in the laboratory test and in the field, such as abrasion, adhesion, fatigue, existence or not of a third body, erosion, corrosion, their combinations, and so on) [3]. Both mechanical and electrochemical methods allows for monitoring and recording *in-situ* and in real-time the following macroscopic quantities as normal or tangential force, relative or rotational displacement, sliding velocity, angular frequency, contact temperature, vibrations of the contacting parts and in relative motion, noise eventually emitted during the test, and the measurements of the electrochemical potential, the corrosion current and/or the impedance of the working materials. In most cases, both wear and corrosion rates can be determined *ex-situ* from a loss of material on one or both contacting materials by surface characterization methods and Faraday's law concepts respectively. Since the chapter does not cover comprehensive information regarding tribocorrosion nor it is the scope of the present work, a state-of-the art and critical reviews on tribocorrosion with specific discussions related to mechanisms, general procedures, and technological aspects may be found elsewhere [1–6], and it is left to the reader to locate further reading sources for that type of information.

The author assumes the reader is familiar with the background of corrosion phenomena and of conventional electrochemical methods in corrosion research. Many textbooks and review articles in the field are available but, for the sake of readability, a reminder of some aspects is given here.

2. Importance of corrosion and tribocorrosion

One of the primary reasons for the importance of corrosion or tribocorrosion lies on global economic losses. The destruction of nearly a quarter of the world's annual steel production is caused by corrosion. An estimate of about 150 million tons of losses per year, i.e. 300 tons per minute! [8]. Corrosion is definitely not limited to steel alloys but affects all sort of materials, namely metals, polymers, and ceramics. Corrosion and wear damage to materials, both directly and indirectly, cost industrialized countries hundreds of billions of dollars annually. For example, wear failures of metals costed the U.S. economy almost $20 billion per year (in 1978) compared to about $80 billion annually for corrosion during the same period [6]. The economic losses, due to friction and wear, related to these costs are estimated to be 6–10% of the Gross National Product (GNP). Wear represent 30% of the causes of dysfunction of the global mechanical engineering systems [2]. A recent study commissioned by the American Federal Highway Administration reveals that the annual direct cost of corrosion was $ 276 billion in 1998, which represents 3.1% of the GNP [9]. Similarly, in the United Kingdom, Japan, Australia and Kuwait, the total annual cost of corrosion was estimated to range between 1 and 5% of each country's GNP [10].

Owing to many different types of expenses involved, in general, estimates of the total cost of corrosion and wear evolve over time and are sometimes difficult and uncertain. There is no doubt, however, that the cost is quite elevated. The direct losses concern replacement of corroded materials and equipment. The indirect losses are related to cost of repair and loss of production, cost of corrosion protection, and prevention. The direct losses are very often lower than indirect costs. For example, it is estimated that the price of repairing or replacing a corroded heat exchanger in a nuclear power plant is insignificant compared to the cost of lost production time. Another important aspect among major influential factors that contribute to corrosion or tribocorrosion relevance is related to reliability, or safety and preservation. Corrosion and wear can compromise the reliability and security of the operating equipment, leading to failure in-service, and at worst disasters, e.g., pressure vessels, metal reactors for toxic chemicals, turbine rotors, nuclear power plants, steering mechanisms for vehicles automobiles, and so on. Further, it requires the rebuilding or replacing the corroded structures and machinery or their components and an additional investment of the following supplies and facilities: metals, energy, water and human efforts to design these metal structures, without mentioning any other resources.

3. Relevance of corrosion and tribocorrosion monitoring

The importance of corrosion and/or tribocorrosion monitoring is that it allows people of interest to study the extent of damage due chemical and/or mechano-chemical attacks and to become aware of the rate at which such damage is progressing so that the necessary measures can be taken to avoid trouble. Generally, the application of continuous and adequate on-line corrosion or tribocorrosion monitoring includes the following advantages: enhanced security, ensuring operational reliability on-time, minimizing process contamination and maximizing product quality, providing a sentinel for equipment integrity, and preventing any further risk related to material or production. The most convenient way to successfully combat corrosion or tribocorrosion impact on materials and structures is (i) to understand the main causes of corrosion and/or wear, (ii) to use all available means to prevent it, (iii) and to implement a continuous improvement approach to corrosion and wear protection.

There are several factors that can come into play in the process of corrosion or tribocorrosion [1–3, 8]. Among the common influential criteria on corrosion, the following main parameters can generally be selected, namely the pH of the medium, the presence of chlorides (and other alkyl halides), the oxidizing power, the pressure, and the temperature. Fundamentally, corrosion depends on the dominating deterioration mechanism of surfaces exposed to chemical environment. The corrosion resistance properties can be characterized directly by the limits of use of the materials, which can be expressed, for example, in terms of maximum temperature in-service or maximum concentration of use. Under normal service conditions, the understanding and control of corrosion are based on the electrochemical interpretation of corrosion phenomena and the consideration of the relative ranking scales of materials in order to select, by successive approaches, the materials best suited to each application of interest [3].

Conversely, a more obvious mechanism of tribocorrosion is the periodic exposure of fresh bare surfaces when sliding friction between surfaces occurs in corrosive liquids or gases [3]. This results in reaction products mainly driven by chemical and electrochemical interactions. The surfaces of the materials are quickly covered by a scale of the reaction product, the oxide in the case of metals and metallic alloys, acting as a protective barrier layer. Often, the thinner the scale, the faster the reaction, and the weaker the protectiveness [3]. Though, tribocorrosion processes are complex, combining both wear and corrosion. Modern research has established a consensus on four main forms of wear, namely, chemical wear (i.e., corrosion and corrosive wear), adhesive wear, abrasive wear or surface fatigue wear [3, 11]. Each process obeys its own laws and, to confuse things, one of the modes of wear acts to affect the others, hence the complexity of corrosive wear [1–3]. As a general rule, there is a combination of competitive wear mechanisms in a dynamic mechanical-chemical contact.

3.1 Choice of technique (instrumentation)

There are many electrochemical and non-electrochemical techniques for the study of corrosion or tribocorrosion and many factors must be taken into account when choosing a technical method. The corrosion rate can be determined by extrapolation of Tafel from a potentiodynamic polarization curve [12, 13]. It can also be determined using the Stern-Geary equation from the polarization resistance derived from a linear polarization experiment or from electrochemical impedance spectroscopy [12, 13]. Furthermore, among the recently developed techniques, those using electrochemical noise analysis as a method of determining the rate of corrosion and estimating the tribocorrosion rate of some passive materials has been shown to be beneficial, although scientists are struggling through the interpretation of some conflicting results [2, 12–15]. Sensitivity to localized corrosion is often evaluated by determining a breakdown potential and a repassivation potential for passivating materials.

3.2 Corrosion forms

Obviously, the most typically known mode of corrosion is the rusting of iron, or iron oxide, and ordinary steel. Contemporary corrosion research has established several forms of corrosion, all which are important to understanding, as the best methods of preventing corrosion depend upon the form of corrosion, as shown in **Figure 2**.

- *Uniform corrosion*: the damage by general or uniform corrosion is fairly predictable; however, the damage caused by localized corrosion is rather unpredictable if no proper monitoring techniques are applied. When left uncontrolled, corrosion will not only cause costly equipment maintenance and replacement

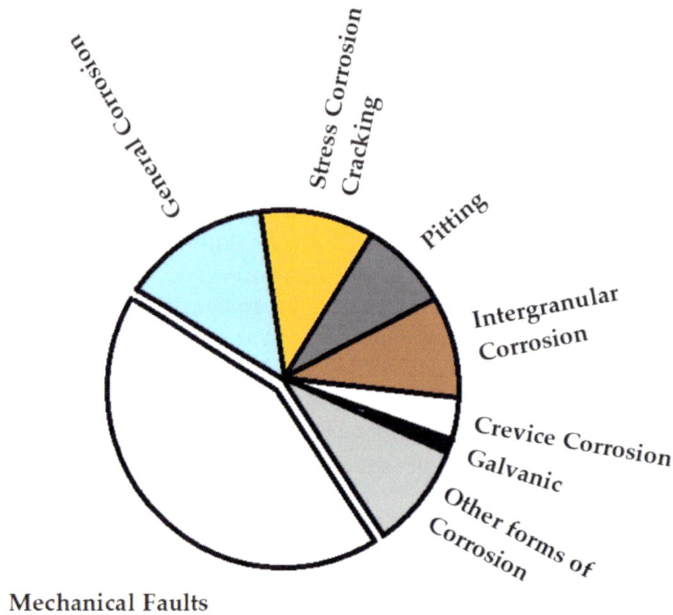

Figure 2.
Forms of corrosion.

but may be responsible for the loss of revenue from unexpected system shutdown and the possibility for hazardous leaks causing major safety issues and possible environmental contamination [16].

- *Pitting* is a form of localized corrosion that occurs when the thin passive protective film on most corrosion-resistant engineering alloys is compromised. The rate of attack at pits can be extremely high so pitting can lead to perforation of a structure or initiation of a crack [17].

- Localized corrosion often initiates at an occluded region where the environment has limited access. This form of corrosion, so-called *crevice corrosion*, is extremely important in fastened structures which often contain many crevices [17].

- Localized corrosion is sometimes observed at grain boundaries when the composition of the grain boundary or region near the grain boundary is different than that in the metal grain. This type of corrosion, so-called *intergranular corrosion* is a severe problem, in particular for stainless steels and aluminum alloys [17].

- Corrosive attack can be localized at one component of a structure made from different metals that are electrically connected as a result of *galvanic corrosion* [17].

- The corrosion of metal alloys often results in preferential reaction of one or more of the alloying elements, so-called *dealloying* [17].

- A common by-product of the corrosion process is hydrogen. Hydrogen can interact with metals in various ways to result in a degradation of properties, primarily mechanical properties (e.g., pipeline gas transportation by H_2S). A generic name for such degradation is *hydrogen damage* [17].

- The mechanical properties of metals can be severely degraded by the combined effects of the environment and an applied stress. *Stress corrosion cracking* is the premature failure of metal structures as a result of these effects [17].

- *Corrosion fatigue* occurs when the applied stress is fluctuating rather than being constant [17].

4. Aspects of corrosion prediction in aqueous media

Although, air in atmosphere is the most omnipresent environment, the aqueous solutions (e.g., atmospheric moisture, acid rain, natural waters, artificial solutions, and so on) are the most common environments associated with corrosion issues. Due to the conditions under which a material, e.g., a metal or a metallic structure is exposed to the environment, corrosion processes usually take place at the metal/environment interface through ionic conduction processes. Corrosion is due to electrochemical reactions involved in that interface and strongly affected by a number of factors, among others, such as the power oxidizing (electrode equilibrium potential), hydrogen ion activity (pH, acidity level of the solution), driving force for metal stability (Gibbs free energy change), rate-determining step reaction or reaction rate (corrosion current), and temperature (Arrhenius oxidation rate or the Pilling-Bedworth ratio).

The prediction of corrosion of a metal or an alloy in environments such as aqueous solutions requires information on the expected state of the metallic alloy (e.g., oxide or non-oxidized metal, etc.) and the rate at which the metallic alloy moves to that state. So first, thermodynamic principles can be applied to determine what processes can occur, and under which conditions the reactions are at an electrochemical equilibrium and, in case of deviation from that equilibrium, in which directions the reactions can proceed and what is the driving force involved. The kinetic laws then describe the reaction rates. These are strongly related to the activation energies of the electrode processes, to the mass transport and to the fundamental properties of the metal/environment interface, such as the resistance of surface films. A general method, namely the mixed potential theory, is implemented for interpreting or predicting the corrosion potential and reaction rates.

In what follows, a very brief reminder of corrosion principles is given.

4.1 Electrochemical nature of metal/electrolyte exchange reaction

The corrosion process is electrochemical in nature. In a more general sense, a corrosion process involving a metal/aqueous solution or any other metal/ionically conducting medium exchange or interaction interface is defined as any process causing the metal, M, to loose one or more of its loosely bound electrons in the metallic state (i.e., oxidation reaction) to generate a solvated cation in the solution, and it is simultaneously balanced, in order to ensure an exact count of the electrons involved, by a reduction reaction by which one or more atoms of a molecule or an ion (anion) of species in solution gain one or more of these electrons (conservation criteria).

The aqueous corrosion sometimes requires that the oxidation product is either an oxide of the metal itself according to (1):

$$2M + O_2 = 2MO \tag{1}$$

or is a metal hydroxide and the reduction product of hydrogen ions in solution is gaseous hydrogen as claimed in the equation (2):

$$M + 2H_2O = M(OH)_2 + H_2 \tag{2}$$

However, it is of course not necessary for an oxide or hydroxide to be formed as an oxidation product of the metal. For example, in the overall reaction of corrosion of iron in a dilute chloridric acid, the ferrous chloride compound is generated as an oxidation product according to the reaction (3),

$$Fe^0{}_{(s)} + HCl^0{}_{(aq)} = FeCl^0{}_{2(s)} + H^0{}_{2(g)} \tag{3}$$

The superscript associated at each reactant or reaction product refers to the valence charge or the oxidation state. The subscript corresponds to the matter phase state for each element of the reaction or compound.

The foregoing overall reaction involves vigorous oxidation and reduction (*viz.* redox) processes. This requires that hydrogen gas is released, more likely by a desorption process, and the solid Fe is oxidized by the acid, forming an ionic bond with the chloride ions. The ferrous ions are removed from the metal lattice and get solvated as they diffuse into the solution. The ferrous chloride compound can be dissociated at normal concentrations.

It is worthy of note that the chloride ions are not actually involved in the redox reaction, this equation (4) can be written in the simplified form as follows:

$$Fe^0{}_{(s)} + 2H^+{}_{(aq)} = Fe^{2+}{}_{(aq)} + H^0{}_{2(g)} \tag{4}$$

It is worthy of note that this also applies to aqueous corrosion involving other acidic media (e.g., orthophosphoric acid, H_2SO_4, HF), and water-based organic acids (e.g., methanoic or ethanoic acids). In each case, no more than hydrogen ion is active, whereas the other ions such as phosphate, sulfate or acetate take no part in the corrosion reaction. Hence, iron reacts with the hydrogen ions of the acid solution to form iron ions (oxidation process) and hydrogen gas (reduction of process). Thus, Eq. (4) can be conveniently divided into two reactions (partial half-cell reactions), namely the oxidation of iron and the reduction of hydrogen ions. This provides the basis for the thermodynamic consideration of the overall corrosion process as stated by equations (5), (6), and (7):

$$Fe^0{}_{(s)} = Fe^{2+}{}_{(aq)} + 2e^- \quad \text{, anodic half-cell reaction, oxidation} \tag{5}$$

$$2e^- + 2H^+{}_{(aq)} = H^0{}_{2(g)} \quad \text{, cathodic half-cell reaction, reduction} \tag{6}$$

$$Fe^0{}_{(s)} + 2H^+{}_{(aq)} = Fe^{2+}{}_{(aq)} + H^0{}_{2(g)} \tag{7}$$

Oxidation or commonly referred to as the anodic reaction is indicated by an increase in the valence charge or electron production. A decrease in the valence charge or an electron consumption represents a reduction or commonly a cathodic reaction.

Equations (5) and (6) outline the two half-cell electrode reactions-both need to occur simultaneously and at the same speed on the metal surface otherwise it would charge spontaneously electrically, which is impossible. This explains one of the fundamental principles of corrosion, namely: the oxidation rate is equal to the reduction rate (in terms of electron production and consumption).

DOI: http://dx.doi.org/10.5772/intechopen.85392

The sites for the oxidation reactions are called *anodes*, and the sites for the reduction reactions are called *cathodes*. Anodes and cathodes can be spatially separated at fixed locations associated with heterogeneities on the electrode surface. Otherwise, the locations of the anodic and cathodic reactions can randomly fluctuate across the electrode surface. The former case results in a localized form of corrosion, such as pitting, crevice corrosion, intergranular corrosion, or galvanic corrosion, and the latter case results in nominally uniform corrosion [17].

The above concept is illustrated in **Figure 3**. An iron atom undergoes a transformation into a ferrous ion and two electrons. These electrons, which remain in the metal, are immediately consumed during the reduction of hydrogen ions. **Figure 3** illustrates the separation of these two processes in space, merely for the sake of transparency. In fact, whether or not they separate or take place at the same surface location, this does not impact the principle of charge conservation. In some cases, the oxidation reaction takes place in a uniform manner on the surface. In other situations, the corrosion reaction occurs at specific locations of the surface (i.e., localize corrosion).

It should be noted that there are several different cathodic reactions encountered repeatedly during metallic corrosion. The most common are:

Hydrogen evolution:	$2H^+_{(aq)} + 2e^- \rightarrow H_{2(g)}$	(8)
Oxygen reduction (in acid solutions):	$O_{2(g)} + 4H^+_{(aq)} + 4e^- \rightarrow 2H_2O_{(aq)}$	(9)
Oxygen reduction (in neutral or basic solutions):	$O_{2(g)} + 2H_2O_{(aq)} + 4e^- \rightarrow 4OH^-_{(aq)}$	(10)
Metal ion reduction:	$M^{n+}_{(aq)} + ne^- \rightarrow M^{(n-1)+}_{(aq)}$	(11)
Metal deposition:	$M^+_{(aq)} + e^- \rightarrow M_{(s)}$	(12)

n is the number of electrons involved in the reaction. The evolution of hydrogen is a common cathodic reaction because acidic media are frequently encountered. Oxygen reduction is very common because any aqueous solution in contact with air is able to produce this reaction. In nearly neutral electrolytes, the reduction of oxygen is normally one of the most universal cathodic reactions. Concentrations of ≈6 ppm dissolved oxygen are usually present in dilute aqueous electrolytes at ambient temperature (25°C). The reduction of metal ions and the deposition of metals are less frequent reactions and are most often encountered in chemical processes.

4.2 Thermodynamic approach to corrosion

Thermodynamic data estimations are meaningful in corrosion field because they can be used to predict the tendency of a metal to corrode in a given environment. Details on thermodynamic principles can be found in a number of manuals [18, 19]. The Atlas of Electrochemical Equilibria in Aqueous Solutions by Marcel Pourbaix provides a complete summary of the application of thermodynamic corrosion for all elements in water [20]. Dr. Pourbaix originally developed "corrosion maps or corrosion road charts" in the form of potential-pH diagrams [20], which enables for displaying the stability of metals (the lowest free energy state) as a function of the activity of hydrogen ions (pH) and equilibrium potential (oxidizing power). So, the diagrams can, in principle, indicate under which conditions of pH and equilibrium potential, an expected corrosion does not occur (immunity region, e.g., pure Fe) or might cause the metal to transform to an ion (metal loss, state of corrosion, e.g., ions of Fe and its compounds) or even indicate whether an oxidized solid should be formed or not (possible passivity, e.g., oxides of Fe). A typical example showing all of these common aspects of metal corrosion is depicted in **Figure 4** for Fe-H_2O system [20].

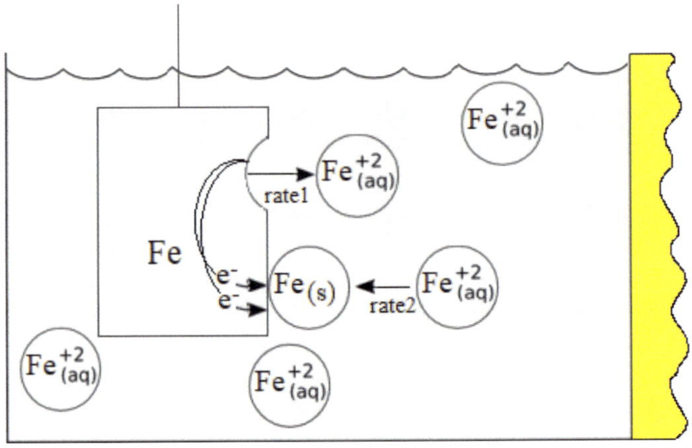

Figure 3.
Reversible iron electrode.

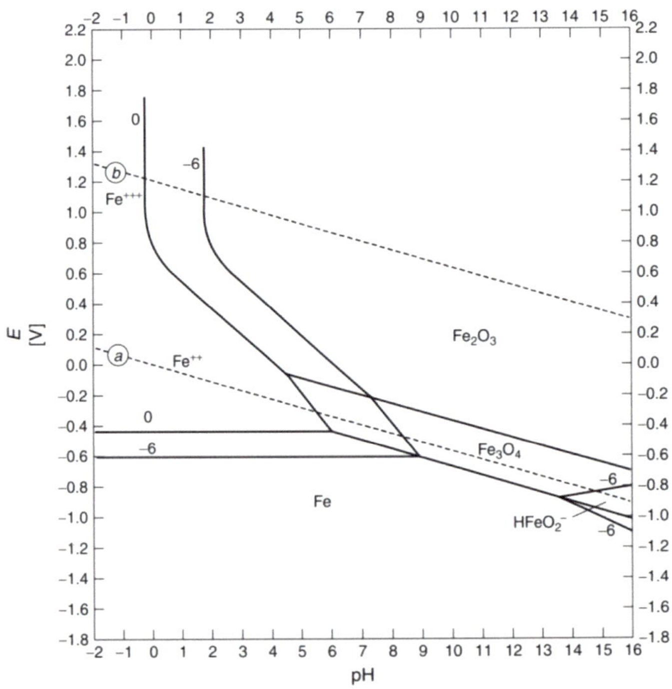

Figure 4.
Pourbaix potential-pH diagram for Fe/H_2O system at 25°C. Reprinted from [20] with permission from NACE international.

4.2.1 Change in Gibbs free energy

A myriad of chemical oxidation-reduction reactions can be performed in an electrochemical cell in which electrons released by the oxidation of species at an electrode (anode) flow through an external conductor (e.g., electrolyte) to a second electrode (cathode) where they are consumed by a reduction reaction [19]. An equivalent ion current flows in the electrolyte which separates the electrodes from the electrochemical cell. In such a device system, the change of chemical energy results from the change of free energy. This change in free energy or commonly termed Gibbs free energy, ΔG, can be defined as a direct measurement of the maximum available electrical energy of a system or in other words it is an indicative measure of the driving force for the reaction to proceed. If the free energy change accompanying the transition of a system from one state to another is negative ($\Delta G < 0$), it indicates a loss of free energy as well as the direction of spontaneous reaction of the system. In particular, if no external force acts on the system, it will tend to transform into its lowest energy state and for a given reaction, it will tend to proceed spontaneously. If the free energy change is positive ($\Delta G > 0$), it indicates that the transition represents an increase in energy, which requires the addition of an extra energy to the system for the reaction to proceed.

The free energy change accompanying an electrochemical reaction can be calculated by the following equation:

$$\Delta G = -nFE \tag{13}$$

where ΔG is the standard free energy in the overall reaction, n the number of electrons involved in the reaction, F the Faraday constant, and E equals the cell potential or the difference between the two standard half-cell potentials. Noting that the term "standard" potential is the potential of a metal or any reactant in contact with its own ions at a concentration equal to the unit activity (i.e., approximately 1 g atom weight per 1 l solution) and at 25°C.

Although Eq. (13) forms the basis for thermodynamic calculations, it is rarely used in studying corrosion phenomena. This is because it is not possible to accurately predict the velocity of a reaction from the change in free energy. It is worth remembering that in most cases, the true magnitude of free energy change is trivial in corrosion applications. The most important factor is the sign of the free energy change for a given reaction, because it indicates whether the reaction is spontaneous or not. Hence, this parameter only reflects the direction of reaction by its sign. Reactions occur exceptionally in the direction that reduces the Gibbs free energy. The more negative the value of ΔG, the greater the tendency for the reaction to proceed. For example, for any overall corrosion reaction of Fe in dilute chloridric acid solution to be thermodynamically possible, it is necessary that the anodic dissolution potential of Fe (i.e., oxidation of Fe) be more active (less noble) than that of the cathodic reaction (i.e., hydrogen evolution). If the change in free energy is positive, the reaction does not occur spontaneously.

4.2.2 Cell potentials

The cell potential of the overall electrochemical or corrosion reaction, as specified above for the example of Fe in a dilute HCl solution, is directly associated with the change in free energy accompanying that reaction via Eq. (13) according to,

$$E = -\frac{\Delta G}{nF} \tag{14}$$

A simple rule can be derived from Eq. (14) to predict the spontaneous direction of any electrochemical reaction. This rule can be clearly stated as: "In any electrochemical reaction, the most negative or active half-cell tends to be oxidized, and the most positive or noble half-cell tends to be reduced". Therefore, the potential is a measure of the reaction (corrosion) tendency. Positive potential corresponds to negative ΔG and hence to spontaneous reaction.

A potential can be assigned to each of the half-cell reactions (anodic oxidation and cathodic reduction). In that respect, the choice of an appropriate reference (half-cell reaction) will therefore allow the measurement of the potential between two electrodes (i.e., working and reference electrodes). An arbitrary half-cell reaction is used as a reference by setting its potential to zero and all other half-cell potentials are calculated with respect to this zero reference. It is common to use the normal or standard hydrogen electrode (SHE) as a universal reference when enumerating the potential of a half-cell [19]. The SHE corresponds to the half-cell reaction (15) given hereafter:

$$2H^{+} + 2e^{-} = H_2 \tag{15}$$

under standard conditions, *viz*. the activity of the protons is unity ($aH^{+} = 1$), and the partial pressure of $H_{2(g)}$ is 1 bar ($p(H_2) = 0.987$ atm). The potential of the SHE is arbitrarily set at zero ($E°_{SHE} = 0$ V), and therefore $\Delta G°$ ($aH^{+} = 1$) is arbitrarily set to 0 as a reference free energy change. The superscript "0" in the SHE potential refers to a standard or reversible potential.

Since it is not possible to manufacture an electrode from hydrogen, an inert electrode is used. Generally, a platinum electrode acts as an inert metal substrate for the electrochemical reaction. Consider an electrochemical cell containing iron and platinum electrodes immersed in a sulfuric acid, both in equilibrium with their reaction product ions and separated by a porous membrane to retard mixing, as illustrated in **Figure 5**. In the reversible divided cell, equilibrium is established between iron and its ions on the iron electrode and hydrogen gas and hydrogen ions on the platinum electrode respectively.

At different locations on the platinum electrode, the hydrogen ions are reduced to gaseous hydrogen and the gaseous hydrogen is oxidized to hydrogen ions, with electron transfer occurring between these spots. It is worthy of note that the platinum electrode does not participate in this reaction but merely serves as a solid interface at which this reaction can occur. Many metals function as reversible hydrogen electrodes (e.g., Hg/Hg_2Cl_2 (Calomel), Ag/AgCl, Hg/HgO, Hg/Hg_2SO_4, Ag/Ag_2SO_4, $Cu/CuSO_4$, etc.); platinum is generally preferred because of its inertia

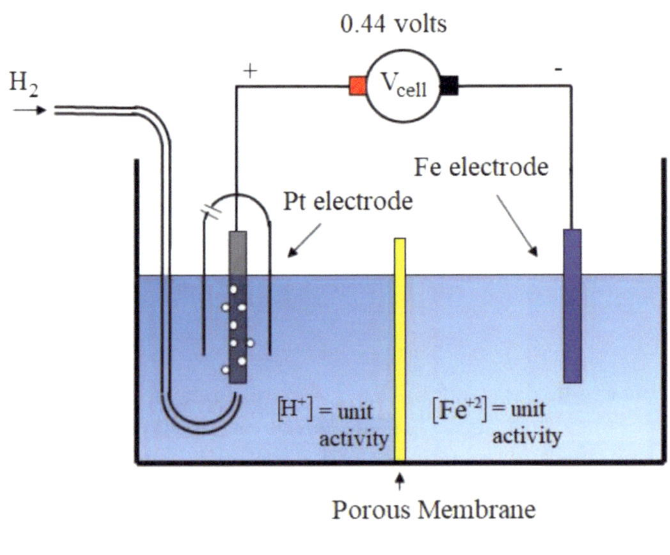

Figure 5.
Electrochemical cell containing reversible iron and hydrogen electrodes (Pt.).

and the ease with which electron transfer occurs on its surface (e.g., properties of catalytic surface exchange or easy current flow rate). As with the other half-cell electrodes, at certain infinitesimal areas on the metal iron surface, iron atoms are oxidized to ferrous ions, and at other spots ferrous ions are reduced to metallic iron. Equilibrium conditions dictate that the rates of the reactions in each half-cell compartment be equal; there is no net change in the system. Noting that the interactions taking place at the metal electrode/electrolyte are quite complex and closely related to electrochemical kinetics and physical phenomena (e.g., electrical double layer, diffusion, etc.).

Consider now the case, under standard conditions, where two dissimilar metals each immersed separately in their own ions in an aqueous medium and electrically connected. Under these conditions, the corrosion control is governed by the determination of the mechanisms involved in the redox reactions. Standard redox potentials can be used to understand the corrosion tendency of metals. A disposition of metals based on their standard potentials has been established and known as standard equilibrium oxidation-reduction potential or standard reversible potential or electromotive force series. The standard potential series of metals is given in **Table 1**. In this series, electrode potential values are invariant, given *vs*. SHE, and attributed to reduction reactions. The same series can be developed on the basis of the oxidation reactions, in which case the potential values will be the same but the "signs" will be reversed.

Redox potentials are very useful for predicting corrosion behavior of metals and upon certain conditions, the tendency of chemical degradation of passivating materials under sliding contacts [22]. Any electrochemical reaction can, therefore, be studied on the basis of the half-cell reaction concept and the foregoing potential standards involving the combination of metals in aqueous solutions with or without any mechanical contact. For example, consider two dissimilar metal pieces, namely silver and zinc, each immersed separately in a dilute acid solution (e.g., H_2SO_4) with their own ions (**Figure 6**) and electrically connected via a potentiostat (i.e., a high-resistance voltmeter). It is necessary that the two electrodes are in equilibrium in their own compartment. The positive terminal of the potentiostat must be connected to the silver electrode, and the negative terminal must be connected to the zinc electrode to have the potentiostat read on scale. According to data in **Table 1**, the standard redox

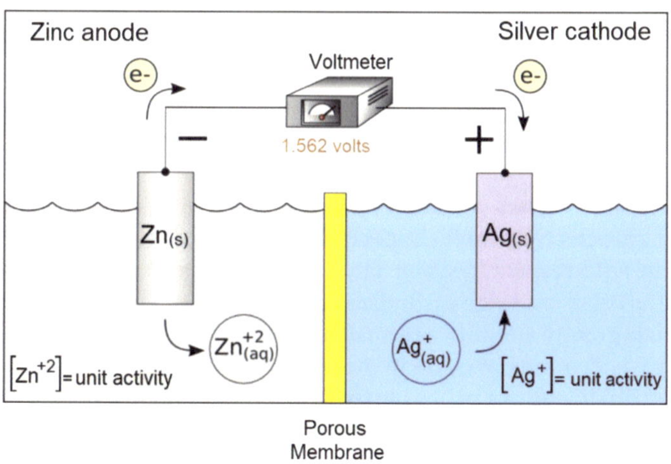

Figure 6.
Electrochemical reversible cell containing silver and zinc in equilibrium with their ions and electrically interconnected via a high-resistance voltmeter.

Electrode red-ox reaction	E° (volts)
$Au = Au^{3+} + 3e$	+1.498
$O_2 + 4H^+ + 4e = 2H_2O$	+1.229
$Pt = Pt^{2+} + 2e$	+1.2
$Pd = Pd^{2+} + 2e$	+0.987
$Ag = Ag^+ + e$	+0.799
$2Hg = Hg_2^+ + 2e$	+0.788
$Fe^{3+} + e = Fe^{2+}$	+0.771
$O_2 + 2H_2O + 4e = 4\,OH^-$	+0.401
$Cu = Cu^{2+} + 2e$	+0.337
$Sn^{4+} + 2e = Sn^{2+}$	+0.15
$2H^+ + 2e = H_2$	0.000
$Pb = Pb^{2+} + 2e$	−0.126
$Sn = Sn^{2+} + 2e$	−0.136
$Ni = Ni^{2+} + 2e$	−0.250
$Co = Co^{2+} + 2e$	−0.277
$Cd = Cd^{2+} + 2e$	−0.403
$Fe = Fe^{2+} + 2e$	−0.440
$Cr = Cr^{3+} + 3e$	−0.744
$Zn = Zn^{2+} + 2e$	−0.763
$Al = Al^{3+} + 3e$	−1.662
$Mg = Mg^{2+} + 2e$	−2.363
$Na = Na^+ + e$	−2.714
$K = K^+ + e$	−2.925
Reproduced from [21] and used with permission.	

Table 1.
*Electromotive force series (Emf or standard potentials redox in volts **vs.** SHE at 25°C).*

potential of silver is +0.799 volts (*vs*. SHE) and that of zinc is −0.763 volts. In this manner, zinc oxidation and silver reduction reactions are spontaneous at +0.763 and +0.799 volts respectively. Their equivalent non-spontaneous reactions occur at −0.763 volts for zinc reduction and at -0.799 volts for the oxidation of silver.

The potentiostat shows a potential difference of approximately 1.562 volts when the silver and zinc electrodes are connected. In this reversible electrochemical cell, zinc is negative with respect to silver. Thus, zinc undergoes anodic oxidation (corrosion) while silver ions undergo cathodic reduction at the silver cathode. The use of redox potentials greatly simplifies the calculation of potential cells and simultaneously allows to determine the trend in metal corrosion.

It follows from the above principle that all metals having a more active (negative) reversible potential than hydrogen will tend to be corroded by acid solutions. Silver, with its higher positive and noble potential, is not corroded in acid media. Though, the presence of dissolved oxygen may alter this latter rule, leading to oxygen reduction (cathodic reaction) and silver oxidation (anodic reaction). **Table 1** indicates that in the presence of oxygen, silver tends to corrode spontaneously.

In that respect, the following conclusions may be drawn from the outcome of the standard redox potential series:

$$Ag + H_2SO_4 = \varnothing \text{ or no reaction } (Ag/Ag^{+} \text{ more positive or noble than } H_2/H^{+}) \tag{16}$$

$$2Ag + 2H_2SO_4 + O_2 = 2AgSO_4 + 2H_2O \ (O_2/H_2O \text{ more positive or noble than } Ag/Ag^{+}) \tag{17}$$

As the reversible potential of a metal becomes more noble, its tendency to corrode in the presence of oxidizing agents decreases.

It is important to note that all the above discussions refer to conditions for corrosion systems at unit activity. Since half-cell potentials change with concentration, Nernst calculations must be made before predicting spontaneous direction at concentrations other than unit activity. To determine the potential of a metal in which the reactants are not at a unit activity, the familiar Nernst equation has been formulated as follows,

$$E = E^0 + \frac{2.303RT}{nF} \log\left[\frac{a_{oxid}}{a_{red}}\right] \tag{18}$$

where E represents the potential of the half-cell reaction, $E°$ the standard potential of the redox reaction. R, T, n, and F are the gas constant, the absolute temperature in kelvin, the number of electrons, and the Faraday's constant respectively. a_{oxid} and a_{red} represent the activities (concentrations) of both oxidized and reduced species respectively.

As indicated in Eq. (18), the half-cell potential becomes more positive as the amount of oxidized species increases. For each tenfold increase in oxidized reactant, the half-cell potential increases by 59 mV for a single electron reaction (n = 1).

In practice, for corrosion to occur, it is not necessary for a metal to be in contact with its own ions. Local anode and cathode zones may exist on the same surface of the metal or alloy, thus acting as an electrical conductor of the electrochemical flow. When such a metal is immersed in an electrolyte, it can no longer equilibrate to experimental thermodynamic scales. Therefore, corrosion is a direct consequence of a new equilibrium potential change, which depends on a myriad of physical and chemical properties of these electrochemical systems (e.g., pH, temperature, oxidizing agents and their power, *IR* drop of solution, etc.). This potential is called corrosion potential or mixed potential, which is far different from the standard potential. In this respect, corrosion will depend on the rate-limited electrode kinetics (e.g., charge transfer rate, mass transport, diffusion, etc.).

4.3 Polarization (overvoltage)-exchange current density

The concept of polarization or overvoltage is very briefly addressed here because of its importance for understanding corrosion reactions and its behavior. It represents a prerequisite for electrochemical methods. The rate of an electrochemical reaction is limited by various physical and chemical factors. Therefore, an electrochemical reaction is said to be polarized or delayed by these environmental factors.

Consider an electrochemical half-reaction for an electrode M at equilibrium:

$$M^{n+} + n e^{-} \leftrightarrow M \tag{19}$$

When a reaction is at equilibrium, it does not necessarily mean that the system is at rest and/or inactive. On the contrary, for a system in equilibrium, the reaction rate in the forward direction is equal to that in the backward or reverse direction. This equal rate at equilibrium is defined as the exchange current density, i_0.

It is worth of note that i_0 is dependent on surface concentrations of the reactants, c_R, and those of the products, c_P, according to: $i_0 \propto c_R^m c_P^n$ m and n are exponents.

As reported in the previous section (Section 4.1), the reaction of hydrogen evolution is very often when a metal corrodes in an acidic environment. Interestingly, the current exchange rate for this cathodic hydrogen evolution reaction strongly depends on the catalytic properties of the metal surface on which the reaction occurs. For example, it ranges from 10^{-12} A cm^{-2} for lead (Pb) to 10^{-3} A cm^{-2} for palladium (Pd) at 25°C in 1 M H_2SO_4 [23]. This variability can have broad implications for the rate of corrosion. In fact, the rate of the reaction can be limited by the rate of the cathodic reaction, which is highly dependent on the current exchange rate of the reaction.

Figure 7 shows the point for an equilibrium of a given reaction at the electrode. It suggests a finite exchange current density at a given potential (e.g., reversible potential E_{rev} or $E°_{M/Mn+}$). Notwithstanding, in electrochemistry, potential and current are interdependent. It is possible to control one and measure the other. Since the former current density involves a "two-way" rate process (forward-backward), the net current density is null. However, at a different potential of E_{rev} (i.e., overvoltage), the half-cell reaction will preferentially proceed in one direction and a net current will be measured.

The overvoltage, $\eta = E\text{-}E_{rev}$, is defined as a potential variation with respect to the reversible potential in the equilibrium of a given electrode reaction. This change in potential is associated with a change in the net reaction rate, and it is therefore related to a change in the direction of the reaction taking place, so that the reaction will mainly proceed in one direction according to the reaction rate. The current is anodic (oxidizing or positive) for a potential change to a value greater than the reversible potential (i.e., anodic polarization) and it is cathodic (reducing or negative) for a potential change to a value below the reversible potential (i.e., cathodic polarization), *cfr* **Figure 7**.

Polarization can be conveniently divided into three different types, activation polarization, concentration polarization, and ohmic potential drop (*iR*).

- Activation polarization refers to an electrochemical process controlled by the reaction sequence at the metal/electrolyte interface (e.g., by possible steps of

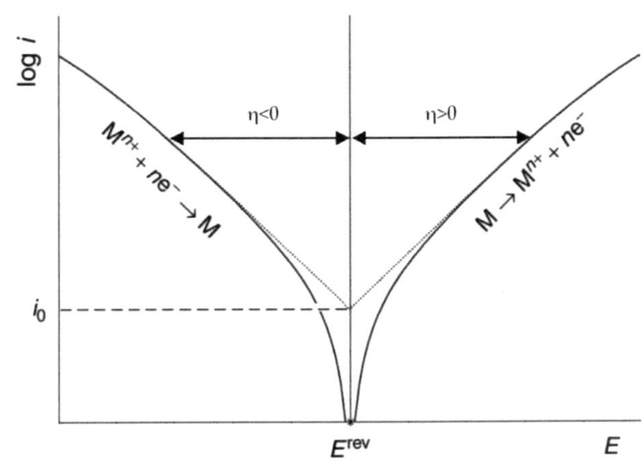

Figure 7.
Illustration of the relation between the current density and the potential for a simple electrochemical reaction under activation control.

adsorption, desorption, and charge transfer processes). It is caused by a slow electrode reaction. In particular, an activation overvoltage or charge transfer overvoltage alludes to an overvoltage resulting from a potential change when the concentrations of the reactants and products at the electrode surface are the same as in the bulk solution. This is because the charge transfer rate at the electrode/electrolyte interface is not infinitely fast. Note that the concentration at the surface of the electrode is equal to the bulk concentration when the mass transport rate is fast compared to the rate of charge transfer.

- Concentration polarization (Diffusion Overpotential) refers to electrochemical reactions controlled by the rate of diffusion in the electrolyte of ions on the surface of the metal.

- iR drop occurs in a portion of the electrolyte surrounding the electrode, or through a reaction product film on the surface of the metal, or in both cases [24]. An ohmic potential drop takes place between the working electrode and the capillary tip of the reference electrode. This contribution to polarization is equal to iR, where i is the current density, and R, equal to l/κ, represents the value in ohms of the resistance path of the solution of length l (cm) and specific conductivity κ (Ω^{-1} cm^{-1}). Therefore, the resistance of an electrolyte of length measuring l (cm) with a cross-section S (cm^2) is equivalent to $l/\kappa S$ (ohms). Hence, the iR drop in volts equals il/κ. The product, iR, decays in parallel with the shutting-off of the current, while the concentration polarization and the activation polarization generally decay at measurable speeds. For instance, for a cathodic protection of steel in seawater (κ = 0.05 Ω^{-1} cm^{-1}), it is necessary to apply a current density of 0.1 A/m^2, which may result in a iR drop correction equal to (10^{-5} V)/(0.05) = 0.2 mV with respect to 1 cm separation of the probe from the cathode. This value is negligible with regards to the critical minimum current density required for convenient cathodic protection. However, in certain soft/fresh waters, where κ can reach the value of 10^{-5} Ω^{-1} cm^{-1}, the corresponding iR drop is equal 1 V/cm.

Noting that the concentration polarization decreases with stirring, whereas the activation polarization and iR drop are not affected significantly.

5. Quantitative measurements of corrosion and tribocorrosion rates

5.1 Corrosion rate

The foregoing discussion on the thermodynamics of corrosion under equilibrium conditions yields information on the driving force of the corrosion process. Since corroding systems are not always at equilibrium, therefore thermodynamic calculations cannot be applied. Notwithstanding that the information on the corrosion tendency accessible from thermodynamic calculations is important and useful. However, most scientific and technical aspects in the field of corrosion focus on the knowledge and the reduction of corrosion rate. Thermodynamics does not address the rate of corrosion.

Metals and non-metals are compared on the basis of their corrosion resistance. For these comparisons to be meaningful, the attack rate of each material must be expressed quantitatively. Corrosion rates can be given in a number of different units using different measures of material loss. A convenient way to determine the rate of corrosion is by the immersion of a sample in a corrosive environment for a period of time and measure the weight loss during that time [25]. The weight loss should be normalized to the exposed surface area of the sample in order to determine its rate of corrosion. Therefore, among the set of units adapted to the corrosion rate is the weight loss per unit area per unit of time; for example, mg cm^{-2} s^{-1}. It is also common to use the percentage of weight loss, milligrams per square centimeter per day and grams per square inch per hour. Though, these do not express the corrosion resistance in terms of penetration. From a technical point of view, the penetration rate, or the thinning of a structural part, can be used to predict the lifetime of a given component. An easy way is to divide this measurement by weight loss by the density of the corroding material to obtain a corrosion rate in units of thickness lost per unit of time. The term mils per year is the most desirable way to express corrosion rates (mpy) or mm $yr.^{-1}$. This expression can be easily calculated from the weight loss of the metal sample during the corrosion test by the formula (20), given below:

$$r = \frac{CM}{\rho nF} i_{corr} \tag{20}$$

where r is the rate of material loss (mm $yr.^{-1}$), M the molecular weight of the corroding metal (g mol^{-1}), ρ the density of metal (g cm^{-3}), i_{corr} the corrosion current density (A cm^{-2}), and C a constant to change the units of thickness and time.

5.2 Tribocorrosion rate

Wear is an unavoidable and a potentially serious problem in all areas of engineering [3]. Designers and engineers who have to make optimal decisions in situations where tribocorrosion considerations are significant, need to know "how long will a component last?". To solve this question, numerous models have been developed so far to distinguish this material loss due to tribocorrosion [3]. These models usually correlate a wear volume or a wear rate with physical and geometrical quantities. Various expressions have since been attributed to this material loss, of which the material loss can be defined in terms of weight, volume, surface, depth, width or even charge density or current density, per unit hardness, per unit frictional dissipated energy (work due to the tangential force), per unit input energy (work due to the normal force), or even per unit sliding distance, or sliding time, sliding frequency, contact frequency, etc. [3]. It becomes readily understandable of the complexity of comparing results between the various wear data published so far. It is expected then that the terminology in this field is rather uncertain, and it will remain so for a certain time, hence the need for a specific standardization, despite some recent progress made in this area [3, 26].

One of the earlier attempts to predict the wear rate or wear volume loss of a material in sliding contact is the commonly Archard wear criterion [27] used during the second half of the 20th century. That criterion is usually expressed as follows [3],

$$W_v = k\frac{F_N}{H} = kA_r \tag{21}$$

where W_v and W_r represent the volumetric loss (assigned as total volume of wear debris produced), and the wear rate (usually expressed per unit sliding distance), respectively. k is the dimensionless Archard wear coefficient, A_r the real area of contact, F_N the applied normal load, and H the hardness of the worn material.

This equation was originally used for the case of adhesive wear [28, 29], then it was extended to more cases including that of tribocorrosion [3]. This is because the k parameter in Eq. (21) exclusively remained the only flexible parameter consistent with the case to which the wear may originate [3]. For example, unidirectional sliding of mild steel against mild steel without any lubricant has a k of 10^{-2}, whereas, for stellite sliding against tool steel, k is 10^{-5} [3, 30]. Even more confusing is that, according to literature, the Archard wear coefficient can vary by two orders of magnitude for the same couple of materials just due to a slight change in load or speed [31]. These findings should be taken with precautions in view of the number of empirical error cases reported with respect to the wear reproducibility and validation of test methods [3]. Friction and wear properties are often considered as subjects of poor accuracy in comparison with materials intrinsic properties. Indeed, comparative round robin studies on the topic have shown that the reproducibility of wear derived from different inter-laboratories with the same material pairing was often very poor [32]. Using the wear track width, the scattering was roughly 50% whereas the scatter in the wear coefficient was over three orders of magnitude. No clear correlation was found between a single and constant parameter (type of tribometer, normal force, and sliding velocity) and the wear rates measured in inter-laboratories [3]. Interestingly, a good convergence was found between the wear volume loss and the energy dissipated in the tribo-contact zone [33–37]. This can readily be explained by the fact that the dissipation of frictional energy is one, among others, of the main causes of tribo-electrochemistry, playing an essential contributing role in wear mechanism, in this case entailing an acceleration (e.g., chemical wear rates) or modification of tribo-chemical reactions [3]. The yielded frictional heat between interacting surfaces leads to a stationary rise in temperature at surface contact asperities and flashes. Furthermore, such frictional energy can take the form of high quantum excitations with short lifetime of surface and bulk sites due to the mechano-chemical forces involved during the sliding process. Those excitations are also responsible for the occurrence of triboluminescence and triboelectricity [3].

Tribologists nowadays are seeking for an agreement due to the fact that there is an unavoidably need to address more fundamental research towards the establishment of an original formulation or a universal methodology to define a "wear criterion" in order to better understand the complexity of the wear process in a tribocorrosion test [3]. Although, this aim has not yet been achieved, a fair amount of progress has been made on this matter-oriented approach. This remains so far valid only as part of the case-by-case study. To conclude, research must focus on establishing a functional-mechanistic based approach that emphasizes the nature of the dependence of the mechano-chemical wear rate (output) on the energetic aspect of sliding friction, the electrochemical aspect of the exposure of bare metal surface, and the transformation of the subsurface material (input). This usually should incorporate materials properties, and behavior. If it does, this could be very useful to help solving the issues and struggling difficulties encountered in the specific field. This will lead to a better improvement of the reliability life of selected material and design technologies when adopted in specific mechanical articulations and under aggressive environments. Further, this could predict materials performance in an environment where tribocorrosion plays a significant role [3].

In tribocorrosion phenomena, where tribological contacts are exposed to corrosive environments, such as aqueous lubricants, the contact materials are subject to both mechanical, and chemical/electrochemical solicitations, which contribute to material removal from sliding surfaces [3]. The rate of material degradation/removal cannot be predicted simply by adding the wear rate in absence of corrosion to the corrosion rate in absence of wear. The reason is that corrosion and wear do not proceed independently and synergistic effects usually (but not always) result in accelerated material degradation (tribocorrosion) [3]. In that respect, theoretical models have been developed so far with respect to mechanical, chemical, and electrochemical factors and their mutual interactions, and which can be tested under well-controlled experimental conditions. In general, modeling has followed either an empirical or a mechanistic approach [3]. The empirical approach is based on the independent measurement of material loss due to wear and corrosion. These parameters are summed up and compared to the material loss due to tribocorrosion. The difference between the two is termed synergy (ΔW_{syn}). A general equation for this approach is of the form [7, 38, 39],

$$W_{tot} = W_{mec} + W_{corr} + \Delta W_{syn} \quad (22)$$

where W_{mec} represents the material loss due to wear measured in the absence of corrosion, and W_{corr} is the material loss due to corrosion only without any influence of mechanical wear.

Although, the empirical approach is technically feasible which allows for the ranking and the performance of materials based on their resistance to tribocorrosion in engineering systems, it is still time-consuming, quite economically not justifiable in the long-term, and furthermore, it integrates a synergy term, which has no physical meaning [3].

The advantage of a mechanistic approach is that it leads for a better understanding of the physical processes involved in tribocorrosion by incorporating the notion of synergism into the mechanical and electrochemical terms [3]. Many factors can be responsible for the mutual dependence of mechanical and chemical material removal in a tribocorrosion system [3]. For example, local abrasion of the passive film can lead to wear accelerated corrosion due to rapid dissolution of the locally depassivated metal surface, followed by repassivation [40]. The abrasive action of hard oxide particles formed by corrosion can accelerate the mechanical metal removal by wear [41]. The plastic deformation of the surface layer of a rubbing metal can lead to a transfer of material to the opposite body resulting in a reduction of the corrosive wear rate [42].

Therefore, it is important to distinguish material loss due to chemical or electrochemical oxidation (i.e., wear accelerated corrosion) from material removed due to mechanical wear (i.e., mechanical material removal from the sliding contact) [3]. The former arises from the fact that an asperity sliding on a material surface produces a fresh wear track zone of clean bare material (i.e., metal), which is usually more susceptible to corrosion than the same surface subjected to free corrosion under no mechanical plastic contact or sliding conditions [3]. The effect of repeated sliding may cause the removal of metal particles by asperities burrowing beneath the surface [42, 43].

Therefore, the overall wear volume due to tribocorrosion, W_{tot}, can be defined as follow:

$$W_{tot} = W_{chem(wac)} + W_{mec} \quad (23)$$

where $W_{chem(wac)}$ is the electrochemical contribution to wear; it is termed wear accelerated corrosion and it reflects the material loss due to corrosion in the presence of wear. W_{mec} is the mechanical wear, and it reveals the material loss due to wear in the presence of corrosion, and which can be related to processes as that for the formation-ejection of oxide debris, oxide layers or any corrosion products, and plastically detached metal [3].

W_{tot} can be determined by measuring the volume of the wear scar post-experiment using, for instance, a laser non-contact profilometry or by on-line measurement of the rate of moving down of the counter-body (e.g., a pin) on the surface wear track during sliding. The latter method has the advantage of recording an instantaneous wear rate, but it would only be applicable if no significant amount of solid reaction products (such as third body particles) accumulate in the contact zone during the tribocorrosion experiment [3]. Under potentiostatic control, the electrochemical term ($W_{chem(wac)}$) can barely be related to the anodic corrosion current ($I_{a,tribocorr}$) measured under mechanical sliding wear (occasionally by subtracting the background current) using Faraday's law. The amount of anodically oxidized metal under such conditions is calculated as follows [14]:

$$W_{chem(wac)} = \frac{Mq}{\rho n F} \tag{24}$$

where $W_{chem(wac)}$ is the volume of the metal transformed by anodic oxidation in a tribo-electrochemical test. q represents the electric charge produced, which results mainly from the integration of the measured current $I_{a,tribocorr}$ under sliding wear conditions over the duration of the tribo-electrochemical test. M, n, and F are the molar mass of the metal, the valence for the anodic oxidation reaction, and the Faraday's constant respectively. ρ is the density of the metal.

The foregoing equation is credible and independent of whether the anodic oxidation leads to the formation of dissolved metal ions or solid reaction products, such as oxide films [3].

It is worthwhile to note that few assumptions must be met in order for Eq. (24) to be used [3, 40], namely:

- The measured current must be equal to the anodic partial current for metal oxidation, which means that cathodic partial currents due to the reaction of oxidizing agents must be negligible. This can be performed by anodic polarization into the passive potential region.

- The charge number "z" for the oxidation reaction must be known [40, 42].

The mechanical wear (W_{mec}) is taken as the difference between the total wear volume W_{tot} and the chemical wear volume $W_{chem(wac)}$ determined from the electric charge.

Author details

Abdenacer Berradja
MTM Department, K.U. Leuven, Leuven, Belgium

*Address all correspondence to: a.berradja@gmail.com

References

[1] Landolt D, Mischler S. Tribocorrosion of passive metals and coatings. Oxford, Philadelphia: Woodhead Publishing Limited; 2011. ISBN: 978-1-84569-966-6

[2] Celis JP, Ponthiaux P, editors. Testing tribocorrosion of passivating materials supporting research and industrial innovation. In: Handbook EFC 62. Leeds, UK: Maney Pub; 2012. ISBN: 9781907625202

[3] Berradja A. Metallic glasses for triboelectrochemistry systems, Chap. 5. In: Huang H, editor. Metallic Glasses-Properties and Processing. London, UK: Intech Open; 2018. ISBN: 978-953-51-6208-7

[4] Uhlig HH. Mechanism of fretting corrosion. Journal of Applied Mechanics. 1954;**21**:401-407

[5] Waterhouse RB. Fretting Corrosion. Oxford: Pergamon; 1972. pp. 182-196

[6] Matsumura M, editor. Erosion-Corrosion: An Introduction to Flow Induced Macro-Cell Corrosion. U.A.E., Japan: Bentham Science Publishers; 2012. DOI: 10.2174/97816080535131120101. ISBN: 978-1-60805-497-8

[7] Madsen BW. Measurement of erosion-corrosion synergism with a slurry wear test apparatus. Wear. 1988;**123**:127-142

[8] Landolt D. Corrosion and Surface Chemistry of Metals. Boca Raton, FL: EFPL Press, Taylor and Francis Group, LLC; 2007, Chap. 10. pp. 415-460

[9] Koch GH, Brongers MPH, Thompson NG, Virmani YP, Payer JH. Corrosion Costs and Preventive Strategies in the United States, Supplement to Materials Performance, July 2002. Report No. FHWA-RD-01-156; Federal Highway Admin., McLean, VA; 2002

[10] Kruger J. Cost of metallic corrosion. In: Revie RW, editor. Uhlig's Corrosion Handbook. 2nd ed. New York: Wiley; 2000. pp. 3-10

[11] Burwell JT. Survey of possible wear mechanisms. Wear. 1957;**1**:119-141

[12] Frankel GS. Electrochemical techniques in corrosion: Status, limitations, and needs. Journal of ASTM International. 2008;**5**:101241. DOI: 10.1520/JAI101241

[13] Frankel GS, Rohwerder M. Electrochemical techniques for corrosion. In: Encyclopedia of Electrochemistry. Weinheim Germany: Wiley VCH; 2007

[14] Berradja A, Déforge D, Nogueira RP, Ponthiaux P, Wenger F, Celis JP. An electrochemical noise study of tribocorrosion processes of AISI 304L in Cl^- and SO_4^{2-} media. Journal of Physics D: Applied Physics. 2006;**39**:3184-3192

[15] Wu P-Q, Celis J-P. Electrochemical noise measurements on stainless steel during corrosion-wear in sliding contacts. Wear. 2004;**256**:480-490

[16] Trenthewey KR, Camberlain J. Corrosion for Science and Engineering. 2nd ed. United States: Longman Group Limited; 1995. ISBN: 978-0582238695

[17] Frankel GS, Landolt D. Fundamentals of corrosion. In: Encyclopedia of Electrochemistry. Weinheim, Germany: Wiley-VCH Verlag GmbH & Co. KGaA; 2007. ISBN: 9783527302505

[18] Roberge PR. Handbook of Corrosion Engineering. New York: McGraw-Hill; 2000. p. 1140. ISBN: 978007-076516-2

[19] Frankel GS, Landolt D. Thermodynamics of electrolytic corrosion. In: Encyclopedia of Electrochemistry. Weinheim, Germany:

Wiley-VCH Verlag GmbH & Co. KGaA; 2007. ISBN: 9783527302505. DOI: 10.1002/9783527610426

[20] Pourbaix M. Atlas of Electrochemical Equilibria in Aqueous Solutions. Houston, TX: NACE International; 1974

[21] DeBethune AJ, Loud NAS. Standard Aqueous Electrode Potentials and Temperature Coefficients at 25°C, Clifford A. Skokie, Illinois: Hempel; 1964

[22] Celis J-P, Ponthiaux P, Wenger F. Tribo-corrosion of materials: Interplay between chemical, electrochemical, and mechanical reactivity of surfaces. Wear. 2006;**261**(9):939-946

[23] Bockris JO'M, Reddy AKN. Modern Electrochemistry. New York: Plenum Press; 1970

[24] Revie RW, Uhlig HH. Kinetics: Polarization and corrosion rates. In: Corrosion and Corrosion Control. 4th ed. Hoboken, New Jersey: John Wiley & Sons, Inc.; 2008. ISBN: 9780470277270

[25] Frankel GS, Landolt D. Kinetics of electrolytic corrosion. In: Encyclopedia of Electrochemistry. Weinheim, Germany: Wiley-VCH Verlag GmbH & Co. KGaA.; 2007. ISBN: 9783527302505

[26] Diomidis N, Celis JP, Ponthiaux P, Wenger F. A methodology for the assessment of the tribocorrosion of passivating metallic materials. Lubrication Science. 2009;**21**(2):53-67

[27] Archard JF. Contact and rubbing of flat surface. Journal of Applied Physics. 1953;**24**(8):981-988

[28] Rabinowicz E. Friction and Wear of Materials. 2nd ed. New York: John Wiley & Sons, Inc.; 1995. ISSN: 0021-8936

[29] Moore MA. Abrasive wear. In: Rigney DA, editor. Fundamentals of Friction and Wear. Metals Park, OH: American Society of Metals; 1980. p. 73

[30] Quinn TFJ. Review of oxidational wear, Part I. Tribology International. 1983;**16**(5):257-270

[31] Welsh NC. The dry wear of steel. Part I. Philosophical Transactions of the Royal Society A. 1965;**A257**:31-50

[32] Czichos H, Beeker S, Lexow J. International multilaboratory sliding wear tests with ceramics and steel (VAMAS 2nd round robin). Wear. 1989;**135**:171

[33] Berradja A, Bratu F, Benea L, Willems G, Celis J-P. Effect of sliding wear on tribocorrosion behaviour of stainless steels in a Ringer's solution. Wear. 2006;**261**:987-993

[34] Mohrbacher H, Celis JP, Roos JR. Laboratory testing of displacement and load induced fretting. Tribology International. 1995;**28**:269-278

[35] Huq Z, Celis J-P. Expressing wear rate in sliding contacts based on dissipated energy. Wear. 2002;**252**:375-383

[36] Fouvry S, Kapsa P, Vincent L. Quantification of fretting damage. Wear. 1996;**200**:186-205

[37] Matveevsky RM. The critical temperature of oils with point and line contact machines. Transactions ASTM. 1965;**87**:754

[38] Watson SW, Friedersdorf FJ, Madsen BW, Cramer SD. Methods of measuring wear-corrosion synergism. Wear. 1995;**181-183**:476-484

[39] ASTM Standard: G119. Standard guide for determining amount of synergism between wear and corrosion. In: Annual Book of ASTM Standards Vol. 03.02. Wear and Erosion, Metal Corrosion. West Conshohocken, PA: ASTM; 2001

[40] Mischler S, Debaud S, Landolt D. Wear-accelerated corrosion of passive metals in tribocorrosion systems. Journal of the Electrochemical Society. 1998;**145**:750-758

[41] Stachowiak GW, Batchelor AW. Corrosive and Oxidative Wear, Chap. 13. In: Tribology Series. Vol. 24. 1993. pp. 637-656. ISBN: 978-0-444-89235-5

[42] Zambelli G, Vincent L. Matériaux et Contact, Une Approche Tribologique. Lausanne: Press Polytechniques et Universitaires Romandes; 1998

[43] Ponthiaux P, Wenger F, Drees D, Celis JP. Electrochemical techniques for studying tribocorrosion processes. Wear. 2004;**256**:459-468

Chapter 9

Electrochemical Techniques for Corrosion and Tribocorrosion Monitoring: Methods for the Assessment of Corrosion Rates

Abdenacer Berradja

Abstract

The corrosion mechanism taking place in an aqueous phase with or without mechanical contact is electrochemical in nature. The electrochemical signal is one of the primary sources of information that relates to behavior in potential, current, and electrical charge of a corroding electrode. It arises from processes that cause corrosion and other electrochemical reactions. In a sliding contact in an ionic electrolyte medium, electrochemistry is more likely to interfere with the tribological behavior of tribocorrosion systems. In recent years, attempts by researchers have been made to control the material loss by electrochemical methods for various engineering systems. Applied online for monitoring in-situ uniform, localized, galvanic, or more forms of corrosion, such techniques are very convenient means to measure corrosion rate of materials. Such methods can also be used in different ways to evaluate their ability to protect materials (as inhibitors, protective layers, coatings). In this chapter, theoretical and experimental applications, fundamental aspects, limits of the electrochemical techniques for corrosion, and tribocorrosion monitoring are presented. Standards developed, so far, by various standardization organizations are reported. Fundamentals of traditional and advanced corrosion methods are described, focusing on their advantages, i.e. sensitivity to low corrosion rates, short experimental duration, and well-established theoretical understanding.

Keywords: corrosion methods, tribo-electrochemistry, tribocorrosion, electrochemical impedance spectroscopy, electrochemical noise analysis, linear polarization resistance, Tafel extrapolation method

1. Application of electrochemical techniques for determining corrosion rates

In the section below, practical examples are described of how a number of electrochemical techniques could be used to forecast corrosion or tribocorrosion behavior in practical case studies. The focus is on laboratory tests for rapid corrosion or tribocorrosion tests. The examples do not provide bit-by-bit procedures for screening most or all potentialities. Also, the discussion is not about how to set up and conduct electrochemical corrosion or tribocorrosion experiments. Such information can be readily found in instruction guidelines manual or standard references [1–17]. The accent is put on the interest and validity of combination techniques to provide a better understanding of the corrosion process and more reliable predictions.

1.1 Linear polarization resistance (LPR)

The concept of "polarization resistance" has presumably been initiated by Bonhoeffer and Jena in 1951 [18]—a subsequent to Wagner and Traud's works [19, 20]. In their study of the electrochemical behavior of iron samples of different carbon contents, they found that the slope of the polarization curve, i.e., the rate of potential change E with external current i, at the corrosion potential (or open-circuit potential of a mixed electrode), was low for some iron samples and large for others. Defining this slope as "polarization resistance," R_P, as a result of Lange's suggestion, it was found that there was an unambiguous correlation between the polarization resistance and the corrosion rate, whereas no correlation was found between the carbon content and the rate of corrosion.

Subsequently, Stern and Geary [21] were the first authors to theoretically establish a linear relationship between the polarization resistance and the corrosion rate based on the kinetics of electrochemical reactions (i.e., corrosion current at open-circuit conditions) and the concept of mixed potential theory, first formulated by Wagner and Traud in 1938 (i.e., parameters of the cathodic and anodic E/i relations) [19]. The advantages and limitations of their method have been discussed in a series of published articles [19, 21, 22], and the linearity of the slope of current-potential plot around the corrosion potential has been verified by experimental evidence, thereby avoiding the problem of large current densities. Their theory has been experimentally supported by other authors [19, 21, 22] for different materials and under a variety of environmental conditions. From the 1960s, plenty of publications [23, 24] reported on the use of the polarization technique, which quickly became one of the main electrochemical techniques routinely adapted to rapid corrosion rate measurements, a condition necessary to its success in industrial monitoring corrosion operations.

For a system in which electrode processes involve a slow reaction step at the electrode surface, the rate of reaction is limited by activation overvoltage; the relationship between the reaction rate, or net current density i, and the driving force for the reaction, or potential E, is given by the Butler-Volmer equation. This equation relates i, for a single electrode process, such as Eq. (1) to E by the formula (2),

$$Fe \leftrightarrow Fe^{2+} + 2e^{-} \quad (1)$$

$$\begin{aligned} i &= i_0\left[\exp\left(\frac{\alpha nF\eta}{RT}\right) - \exp\left(\frac{(1-\alpha)nF\eta}{RT}\right)\right] \\ &= i_0\left[\exp\left(\frac{\alpha nF(E-E_{rev})}{RT}\right) - \exp\left(\frac{(1-\alpha)nF(E-E_{rev})}{RT}\right)\right] \end{aligned} \quad (2)$$

where η is the overpotential, i_0 the exchange current density (rate of either the forward or reverse half-cell reaction) at the equilibrium potential E_{rev}, α the transfer coefficient (usually close to 0.5, but must be between 0 and 1), and n the number of electrons transferred.

The graphical representation of the Butler-Volmer equation, as shown in **Figure 1**, is called the polarization curve.

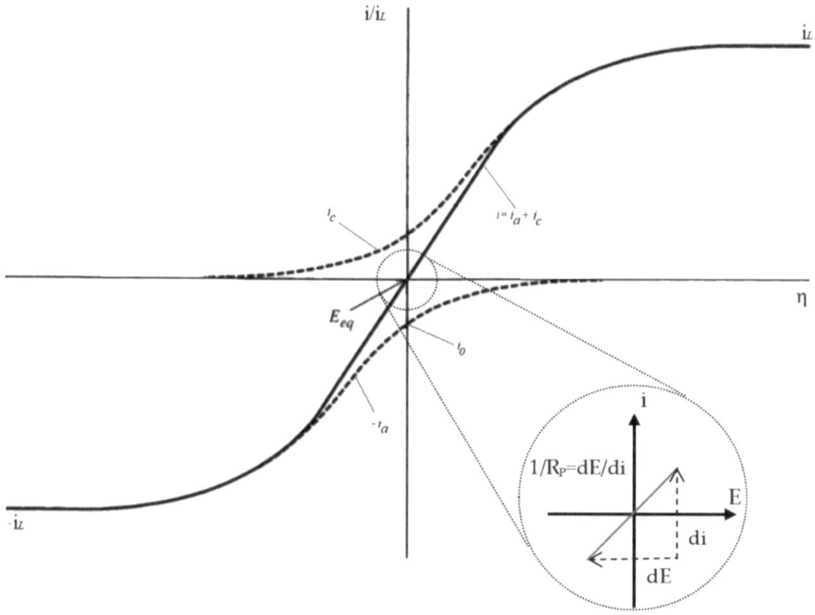

Figure 1.
Current density (i)-overpotential (η) curves for the system O + e ↔ R at 25°C. α = 0.5, $i_{l,a} = -i_{l,c} = i_l$. Partial current densities: i_a, i_c (dashed line), i_l limit current density (horizontal line), and R_P the polarization resistance (circle).

Stern and Geary's theory [21] is based on a simplified corrosion process assuming that only one anodic reaction and one cathodic reaction are involved during the corrosion process. It is therefore inevitable that erroneous results occur when the corrosion process involves more than one anodic or cathodic reaction. To address this problem, Mansfeld and Oldham [25] presented a modification of the Stern-Geary equation by including more than one oxidation and one reduction reaction in a complicated corrosion process. The current-overpotential relationship at electrodes is set by a number of complex physical and chemical phenomena based on experimental conditions. The reactions occurring at the electrode/electrolyte interface are heterogeneous chemical processes that may involve elementary electron-transfer steps (one or more steps) over the electrochemical double layer, ion-transfer, potential independent or chemical steps, etc.

It is well known that the electrochemistry of corroding metals involves two or more half-cell reactions. Suppose there is a simple corrosion system, such as an iron metal (a corroding working electrode) immersed in a sulfuric acid solution, in addition to Eq. (1), the following half-cell reaction (Eq. (3)) also occurs:

$$H^{2+} + 2e^- \leftrightarrow H_2 \tag{3}$$

The dissolution of Fe takes place in the acid electrolyte. At equilibrium, the total anodic rate is equal to the total cathodic rate. In this case, the net rate of either Fe dissolution or hydrogen evolution can be measured at the electrode potential of the steady-state freely corroding condition. This potential refers to the corrosion potential E_{corr}, which lies between the equilibrium potentials of the two individual half-cell reactions. At E_{corr}, the net rate corresponds to the uniform corrosion rate, i_{corr}, at free corrosion condition. In such system, the relationship between the overpotential (η, applied potential minus corrosion potential) and the current (flowing between the working electrode and the auxiliary counter electrode) is governed by the fundamental Butler-Volmer equation given as follows:

$$i = i_0\left[\exp\left(\frac{\alpha nF\eta}{RT}\right) - \exp\left(\frac{(1-\alpha)nF\eta}{RT}\right)\right]$$
$$= i_0\left[\exp\left(\frac{\alpha nF(E-E_{corr})}{RT}\right) - \exp\left(\frac{(1-\alpha)nF(E-E_{corr})}{RT}\right)\right] \quad (4)$$

In **Figure 1**, the linear relationship between the polarization resistance and the corrosion rate can be easily illustrated graphically. In the small region near the corrosion potential, E_{corr}, only a very small perturbation potential, usually less than ± 30 mV (typically ± 10 mV), is applied above or below the corrosion potential, yielding a linear relationship between the overpotential ($\eta = E - E_{corr}$) or the polarization from the corrosion potential and the current. Due to this smooth excitation, the LPR technique is not expected to interfere with corrosion reactions. The slope of that linearized curve ($i - E$) is defined as the polarization resistance, R_P, of a corroding electrode (in ohms cm^{-2} if the current density is plotted or in ohms if the current is plotted), which is mathematically interpreted as

$$R_P = \left(\frac{\partial \eta}{\partial i}\right)_{|E-E_{corr}=0} \quad (5)$$

where i is the current density corresponding to a particular value of E.

The corrosion current, I_{corr}, can be calculated when the overpotential approaches zero and is related to R_P as follows:

$$I_{corr} = \frac{1}{R_p}\cdot\frac{b_a b_c}{2.303(b_a + b_c)} \quad (6)$$

where b_a and b_c are the so-called anodic and cathodic Tafel slopes or Tafel parameters, respectively *(cf. infra)*. The corrosion current density, i_{corr}, can thus be calculated from Eq. (6) if R_P and Tafel constants (b_a and b_c) are known.

ASTM G59 describes an experimental procedure required to carry out polarization resistance measurement [10]. In agreement with this standard, the potential should be scanned from −30 mV to +30 mV of the corrosion potential at a rate of 0.167 mV s^{-1}.

Many of the foregoing determined corrosion key parameters are based on empirical observations. As with any empirical method, due to the high number of factors involved in a corrosion or tribocorrosion system (e.g., environmental changes, temperature, pH, reagent as chloride ions, pressure, specimen geometry, test setup configuration, etc.), it is not uncommon to observe that the values of b_a, b_c, and R_P are influenced by these operational parameters and are therefore subject to change. Of significance, the slope generated from the *i–E* curve around the corrosion potential may not be linear and may or may not be symmetrical in the anodic and cathodic regions. The symmetry of the curve (*i–E*) at the point of equilibrium or at open-circuit potential is obtained only when b_a and b_c are equal. These values are required for computing the corrosion current and are usually determined by the Tafel extrapolation method (*cf. infra*).

It is worthy to note that the measurements of R_p can be derived potentiodynamically or by the method of stepwise potentiostatic polarization or by anodic step pulse method. In the potentiodynamic method, the potential is swept at a constant rate (typically 60 mV/h) from the active (cathodic) direction to the noble (anodic) region passing through the corrosion potential while tracking the current density continuously. More information regarding this method can be found elsewhere [3].

Similarly, in the step pulse method, an applied potential is incremented in steps of ±5 or ± 10 or ± 20 mV, starting from a negative potential moving to a positive potential through the corrosion potential. The value of R_p is determined from the slope of the plot of the potential-current. Prior to the tests, a steady-state corrosion potential is required. The open-circuit potential of the corrosion system is first measured, typically for 1 hour (during which time the corrosion potential of most electrodes is stabilized) or until it reaches a stationary state.

Progress is made through competitive advantages between different measurement techniques, including a rapidity in current measurement (generally rather quickly in a few minutes), where only a lower excitation is required (less than ±30 mV, generally ±10 mV), so that the corrosion rate would not be affected by corrosion reactions, an easy measurement of low corrosion rates (less than 0.1 mil/year (2.5 μm/year), and measurements taken repeatedly, the LPR technique can be considered as a nondestructive technique and used for online corrosion monitoring of uniform corrosion rates useful for the field.

The main drawback of this technique is that the Tafel parameters must be known in advance in order to convert the polarization resistance into the corrosion rate. To tackle this problem, several numerical methods [8, 9, 26–28] have been proposed to obtain both Tafel parameters and corrosion rate from the same polarization measurement in the vicinity of the corrosion rate. Nevertheless, the success is limited since the Tafel parameters thus determined will not be very accurate, which may compromise the nondestructive nature of the LPR technique. Another disadvantage of the LPR method lays in the fact that it will not work properly in low conductive media. Basically, the LPR technique can only be used to determine uniform corrosion rates; it can hardly provide information about localized corrosion.

1.1.1 Illustrative examples of the application of LPR in corrosion and tribocorrosion systems

A modified electrochemical noise technique, namely, electrochemical emission spectroscopy (EES) [29], offers one of the most convincing examples of the application of the LPR technique in tribocorrosion [30]. Indeed, the analysis of noise data in a potential-current plane shows the transposition of the statistical resistance due to electrochemical noise to the resistance due to linear polarization. Noise resistance is often considered equivalent to the polarization resistance, R_P [31–33]. The noise resistance, R_N, calculated using a method proposed by Eden et al. [33], for mild steel passive alloy in 0.05 M H_2SO_4 (corrosion under activation control), is of the order of 48 Ω without any sliding contact. The LPR measured on this mild steel after EES monitoring is shown in **Figure 2a**. The comparative value of R_P obtained by the LPR technique is 50 Ω. The R_N value obtained using the EES technique is therefore very close to the R_P obtained by the LPR technique. Under tribocorrosion conditions (5 N normal force, 10 Hz sliding frequency, 200 μm peak-to-peak displacement amplitude), the plane plot of the potential-current data under steady-state wear-corrosion regime shows a best-fit line through the data points with a positive slope of 54 Ω (see **Figure 2b**), which roughly corresponds to R_N (48 Ω) or R_P (50 Ω in **Figure 2a**). Notwithstanding, no attempt has been made to relate these resistance measurements with the breakdown (i.e., depassivation) or the buildup of any kind of passive film (i.e., repassivation) on the mild steel surface subjected to either a mechanical stimuli (e.g., active-passive wear track zone area or metastable pit area) or in the absence of wear (free corrosion), characteristic phenomena of localized corrosion.

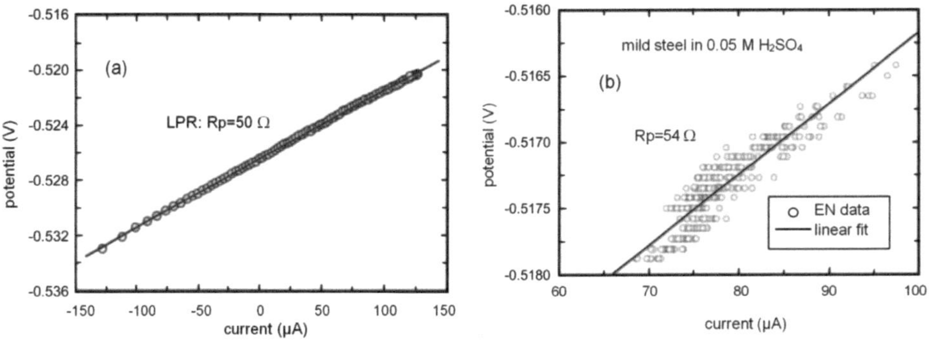

Figure 2.
LPR measurements on the mild steel in 0.05 M H_2SO_4 solution; (a) under free corrosion state, (b) under wear-corrosion steady-state phase. Reproduced from [30] with permission from Wiley Online Library.

1.2 Tafel extrapolation method

In 1905, Julius Tafel [34] presented the experimental relationship between the current, I, and the overpotential, η, during an electrocatalytic test of the reduction reaction of hydrogen (i.e., protons to form molecular hydrogen) on a number of electrode metals such as Hg, Sn, Bi, Au, Cu, Ni, and so on:

$$\eta = a + b \log I \tag{7}$$

where the overpotential η is defined as the difference between the potential of the working electrode, E, and the equilibrium potential.

The existence of a linear relationship between E and $\log I$ has been demonstrated when the electrode is polarized at sufficiently large potentials, and far away from the corrosion potential both in anodic and cathodic directions [34], as can be seen in the polarization curve depicted in **Figure 3**. The portions in which such relationships prevail are called Tafel portions or Tafel regions.

This can be mathematically expressed as

$$\begin{aligned} I &= I_{corr}\left[\exp\left(\frac{2.303\eta}{b_a}\right) - \exp\left(-\frac{2.303\eta}{b_c}\right)\right] \\ &= I_{corr}\left[\exp\left(\frac{2.303(E-E_{corr})}{b_a}\right) - \exp\left(-\frac{2.303(E-E_{corr})}{b_c}\right)\right] \end{aligned} \tag{8}$$

where E_{corr} is the corrosion potential, E the applied potential, η the overpotential (difference between E and E_{corr}), I the current, I_{corr} the corrosion current, and b_a and b_c are the Tafel constants or Tafel parameters derived from $E - \log I$ plots as the anodic and cathodic slopes in the Tafel regions, respectively.

Extrapolating from the Tafel portions of either anodic or cathodic or both, an intersection point is obtained at E_{corr}, from which I_{corr} is readily available from the $\log I$ axis. Therefore, it is possible to obtain simultaneously the corrosion current, I_{corr}, and the Tafel parameters (i.e., b_a and b_c) from this method.

In order to obtain the Tafel portions in the anodic and cathodic regions, the electrode has to be polarized far away from its corrosion potential, e.g., $\pm$250 mV away from E_{corr}. Eq. (8) can be rearranged, as appropriate, to choose one single polarization direction, either anodic or cathodic way.

At sufficiently larger values of η (100 mV $\leq \eta \leq$ 500 mV), in the anodic direction (i.e., $\eta = \eta_a$), Eq. (8) can be rearranged as,

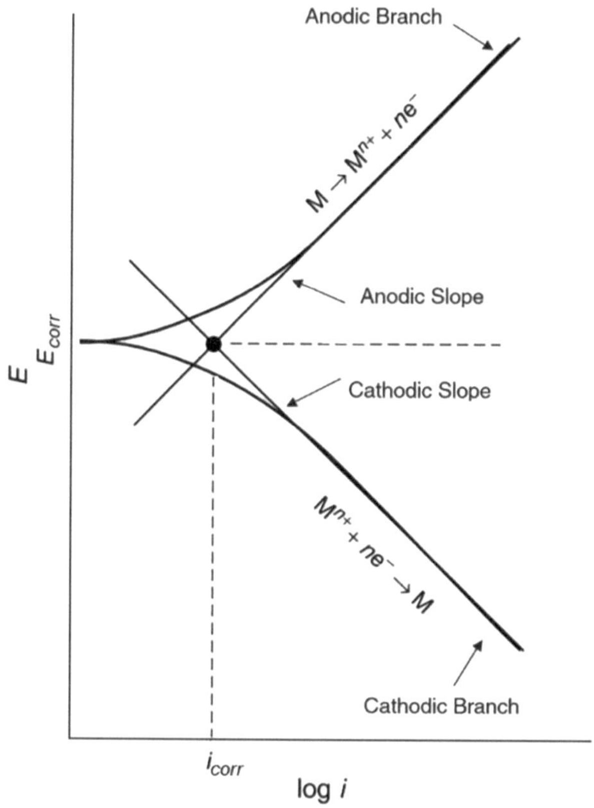

Figure 3.
Electrode kinetics as expressed by the Butler-Volmer equation, plotted in a semilogarithm scale or Tafel plot showing that the corrosion current density can be obtained from the intercept.

$$\eta_a = b_a \log \frac{I}{I_{corr}} \tag{9}$$

Likewise, at sufficiently larger values of η, in the cathodic direction (i.e., $\eta = \eta_c$), Eq. (8) can be modified as,

$$\eta_c = -b_c \log \frac{I}{I_{corr}} \tag{10}$$

The polarization curve can be measured either dynamically or statically (either in the potential-controlled mode or in the current-controlled mode). The dynamic polarization techniques can be carried out relatively fast, but the drawback is that the Tafel parameters are scanning rate dependent. The static polarization techniques may produce better Tafel parameters, but they are very time-consuming.

Tafel extrapolation measurements can be performed either by the potentiodynamic method or by the stepwise potentiostatic polarization method [35]. As in R_P measurements, in both methods, corrosion potential is first measured, typically for 1 h (during which time corrosion potentials of most electrodes are stabilized) or until it stabilizes. After that, the potential step—at increments of ± 25 or ± 50 or ± 100 mV, every 5 min, recording the current at the end of each 5-min period—is applied (potential-step method), or the potential is scanned at a constant rate (typically 0.6 V/h) (potentiodynamic method). In both methods, the experiment is started at the corrosion potential, and the cathodic polarization is first conducted by applying an overpotential of approximately 500 mV or until gas evolution (e.g., hydrogen) occurs at the electrode, at a constant rate of 0.6 V/h. Following, the corrosion potential is measured again (typically for 1 h), and then anodic polarization is conducted by applying an overpotential so that the potential at the end of the anodic polarization reaches +1.6 V versus SCE. Tafel plots are generated by plotting both anodic and cathodic data in a semilog paper as E-log I. From the plot, three values are determined: the anodic Tafel slope, the cathodic Tafel slope, and I_{corr} (from back-extrapolation of both anodic and cathode curves to E_{corr}). The main advantage of this method is that it provides a simple, straightforward method to determine Tafel parameters, namely, b_a and b_c.

The disadvantage of the Tafel technique is that large current densities are often required to generate the complete Tafel plots. The use of large current densities can alter the surface conditions of the specimen (e.g., permanent change or surface damage), thereby distorting the results and increasing complications due to mass transport and uncompensated electrolyte resistance. The measurement of current density over a wide potential range may also distort the results if the adsorption of some species is potential dependent. Since this method applies a large overpotential to the metal surface (e.g., anodic polarization), therefore, the technique is rather destructive and can hardly be used for online corrosion monitoring purposes and in particular in the field. An ASTM G5 standard provides a procedure for constructing an anodic polarization plot [36]. However, it does not supply a method to construct a cathodic polarization plot nor a procedure to determine the corrosion current by the Tafel extrapolation method.

1.3 Corrosion rate determination by electrochemical noise analysis (ENA)

Many of the electrochemical techniques, among those described earlier, measure the electrochemical response of the corrosion system following the application of an external disturbance. In the last 50 decades, an original concept has emerged where it was possible to use the inherent noise of the electrochemical system as a stimulus to measure both potential and current changes [31, 32, 37–43]. Broadly, measured inconsistently in corrosion experiments, the electrochemical noise was first considered an unwanted or undesirable artifact that comes from measuring instruments or pickups from the environment. This is why this misleading name was cast. This sort of noise can be easily observed during corrosion potential measurements because the measured corrosion potential always fluctuates slightly, usually randomly. Random fluctuations result from stochastic processes [44], and, considering each chemical process is stochastic in nature, it generates noise.

Since the pioneer work of Iverson [45], who has reported a relation between the frequency or amplitude of the electrochemical noise and the inhibiting power of the environment for a number of metals and alloys (e.g., aluminum alloys, magnesium, mild steel, etc.), there has been a growing interest toward the measurement of electrochemical noise and its peculiar relationship with localized corrosion [11, 12, 31, 32, 41, 46–51]. In this respect, electrochemical noise measurements obtained from the analysis of corrosion potential or current fluctuations provide a new approach to the study of corrosion processes in reactive environments such as aqueous media or hot aggressive gases or even under the effect of mechanical stimuli, e.g., tribocorrosion. Indeed, mechanical friction of solids in contact with a corrosive environment is likely to generate (i) noise due to stochastic contact between randomly distributed surface asperities and (ii) noise due to the synergy of wear-corrosion processes resulting from the activity of the surfaces and controlled by the response of potential-current transients and the configuration of the wear track area, coordinated by the coupling effects of wear and corrosion in the tribo-electrochemical cell. Among the possibilities offered by the measurement of electrochemical noise sources during an electrochemical or a tribo-electrochemical system, the following can be retained: adsorption–desorption processes, e.g., formation and detachment of gas bubbles; fluctuations in the mass transport rate and in temperature; interfacial nucleation and growth processes; degradation processes due dielectric film disruption; kinetics of atom exchange at the surface sites, e.g., Johnson's noise in the interfacial impedance; and so on.

While multiple case studies on electrochemical noise have been regularly reported in recent years, even greater progress is possible, with the scope for increased breakthrough in science and technology (e.g., novel materials, precision tools on macro-to-nanoscopic scales, availability and intelligent use of these materials and tools, and so on). In particular, the main focus of these investigations is to promptly obtain in situ mechanistic information on the repassivation and breakdown of passive films and to monitor any process associated with confined corrosion and/or tribo- or bio-tribocorrosion [46, 47, 51]. It has been, indeed, suggested that the noise is caused by film breakdown and repassivation processes, and given the dynamic competition between these two processes, pitting will initiate. However, the foundation for using electrochemical noise analysis for determining the corrosion rate of an electrode is still a subject of debate within the scientific community. Indeed, the fundamental approach is not as robust as that of other techniques. On the other hand, the advantage of the noise analysis is that it is not necessary to apply any external polarization and the system is in natural corrosion conditions. This renders the technique as nondestructive and nonintrusive, capable of monitoring basic changes in an electrochemically active system. This makes it particularly suitable for online corrosion monitoring in the laboratory, especially for localized corrosion monitoring, detection of general corrosion, crevice investigation, stress corrosion cracking [12, 52, 53], fretting corrosion, or be used in the assessment of anti-corrosive organic coatings, and other surface inhomogeneity case studies [43, 46, 47]. Several approaches extend the use of electrochemical noise measurements in both pilot plant and field facilities, its use is not merely limited to the foregoing phenomena, but its development is justified especially when measurements are performed in systems with very low conductivity, where, for e.g., the impedance technique fails because of the loss of signal in the high resistance of the solution (*cf. infra*).

1.3.1 Instrumentation for electrochemical noise measurements in corrosion and tribocorrosion systems

Electrochemical noise is a generic term used to describe the naturally occurring fluctuations in potential and current, which is due to spontaneous changes in electrode kinetics and mechanisms [33]. When applied to corrosion studies, electrochemical noise may be redefined as the spontaneous fluctuations observed in potential and current at the free corrosion potential. The electrochemical noise can thus be classified into potential noise and current noise. There are three major possible modes for measuring potential and current noise in a corrosion system, but the most common mode uses two nominally identical working electrodes, WE_1 and WE_2 (WE_1 as the corroding metal and WE_2 as a counter electrode), and a noise-free reference noble electrode, RE [33] (see **Figure 4a**). The current flowing between the two working electrodes is measured by a zero-resistance ammeter (ZRA), and their potential is monitored versus the reference electrode through a voltmeter (V) under free corrosion conditions. The two other leftover modes are two identical working electrodes WE_1 and WE_2 with a bias potential [54] (not shown here) and one WE coupled to a micro-counter electrode (MC, e.g., Pt wire tip) [29, 46, 47, 55]

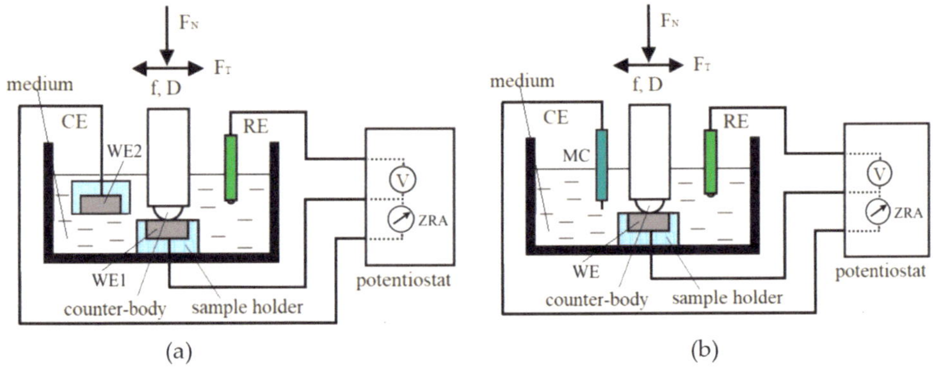

Figure 4.
(a) Schematic view of a tribocorrosion experimental setup. Potential and current are measured on a working electrode (WE_1) sliding against a counterbody ball (unidirectional reciprocating sliding, sphere-on-flat) with respect to a RE *reference electrode (Ag/AgCl (3 M KCl)) via a* V, *high-impedance voltmeter, and* CE (WE_2) *via a zero-resistance ammeter (ZRA), respectively.* F_N, *normal force;* F_T, *tangential force;* f, *sliding frequency;* D, *displacement amplitude. (b) Schematic view of a tribocorrosion experimental setup. Potential and current are measured on a working electrode (WE) sliding against a counterbody ball (unidirectional reciprocating sliding, sphere-on-flat) with respect to a* RE *reference electrode (Ag/AgCl (3 M KCl)) via a* V, *high-impedance voltmeter, and* CE *micro-cathode (MC) via a zero-resistance ammeter (ZRA), respectively.* F_N, *normal force;* F_T, *tangential force;* f, *sliding frequency;* D, *displacement amplitude.*

(see **Figure 4b**). This last mode of electrochemical noise analysis seems to be a promising way to obtain unambiguous estimates of the rate of chemical wear in a tribocorrosion experiment as evidenced by some recent investigations [46, 47, 51] but also to predict the corrosion rate of localized corrosion in a free corroding system [29, 31, 32, 37–45, 47–50, 56].

1.3.2 Electrochemical noise data management process

The overall approach to analyzing noise data is the assessment of mechanistic information from either time-domain analysis, frequency-domain analysis, or both, using statistical methods [44, 57, 58]. If the information in the time-domain records is evident, time-domain analysis is sufficient to distinguish different processes (e.g., different forms of corrosion).

In what follows, one assumes that all various types of noise are excluded from this description, with the exception of the thermal noise. Except for the last noise, all other noise sources can be minimized or eliminated using careful strategy within reasonable limits of materiality. Effective and convenient ways include the removal of unwanted environmental and instrumental noise from the electrochemical noise, e.g., by shielding electrical connections/wires for coupling the electrodes to the experimental apparatus, by using a Faraday cage to exclude electrostatic/electromagnetic influences, even by implementing analogue/digital filters to eliminate systematic noise at frequencies different than the frequency of interest, and so on. Guidelines for the calibration of noise measuring device can be found elsewhere [13, 59].

1.3.3 Noise resistance

The basic quantitative approach is the time-domain analysis of the noise signal. The noise resistance, R_N, is defined as the ratio of the standard deviations of potential/current noise signal time dependent, $\sigma(t)$:

$$R_N = \frac{\sigma_E(t)}{\sigma_I(t)} \tag{11}$$

Eq. (11) implies that in the case where a low-driving force noise produces a high current density noise between the two electrodes (WE_1 and WE_2), the yielding noise resistance will be low. Noise resistance, R_N, has been shown to correlate well with the polarization resistance, R_P, as determined by EIS for certain corrosion systems. This latter being directly related to the corrosion current [14, 29, 60] is using the Stern-Geary equation and Tafel slopes. Notwithstanding, much work has been devoted trying to best match R_N or the normalized R_N (per unit of exposed surface area) to the corrosion resistance or the corrosion rate [5, 14, 31, 32, 37–45, 48–50, 60]. Although signal analysis in the time domain is well established, an approach based on spectral analysis is gaining more and more importance in research laboratories. It consists of transforming the potential and current noise fluctuations recorded in the frequency domain using the Fast Fourier Transform (FFT) method [61].

The frequency range for which the FFT is commonly performed extends from 1 mHz up to 1 Hz. The spectral noise plots are similar to those of impedance plots. The spectral noise resistance, R_{SN}, is given by the ratio of the potential and current FFTs at each frequency, and the limiting value, $R_{SN}{}^0$, can be used as a measure of the corrosion resistance:

$$R_{SN}(f) = \left(\frac{E_{FFT}(f)}{I_{FFT}(f)}\right) \tag{12}$$

The *log–log* plot of R_{SN} versus f is similar to the impedance plot, and the spectral noise resistance limit $R_{SN}{}^0$ is given by

$$R^0_{SN} = \lim_{f \to 0} R_{SN}(f) \tag{13}$$

Another approach would be to examine the spectral noise response in terms of power spectral densities (PSD). These latter are calculated from the FFT or using the maximum entropy method (MEM) [62]. R_{SN} is determined from the PSDs by the relation (14):

$$R_{SN}(f) = \left(\frac{E_{PSD}(f)}{I_{PSD}(f)}\right)^{1/2} \tag{14}$$

It has been shown that the use of a single data set of potential and current noise [32] would yield identical values of R_{SN} as calculated by either Eqs. (13) or (14). In some cases, $R_{SN}{}^0$ is bound to R_N or R_P as,

$$R_N = R_{SN}(0) = R_P \tag{15}$$

effective if the impedance of the two electrodes is identical and much higher than the resistance of the test solution between them [5, 32, 63]. Experiments have validated this relationship for several systems [5, 31, 63]. Nonetheless, there is no agreement on the fundamental basis for the relationship between noise resistance and corrosion rate.

1.3.4 Illustrative examples of the application of electrochemical noise in tribocorrosion systems

Investigations into electrochemical kinetics make common point research between tribocorrosion and corrosion. The study of localized phenomena of depassivation and repassivation is essential to understand the mechanisms of corrosion-wear as well as to reduce the material loss. The possibility of using the electrochemical noise detection technique as a promising tool to study the electrochemical properties of well-controlled damaged surfaces has been widely considered due to its nondestructive nature and its potential in online corrosion monitoring applications. Time-spatially resolved measurements should provide more reliable data on the electrochemical part of tribocorrosion. The noise analysis in relation to depassivation-repassivation events randomly distributed in time and space can be traced back to Oltra et al. [64]. The power spectral density (PSD) of the noise under the impact of the jet particles was related to the Fourier transform of individual repassivation transients obeying a Poisson distribution. Later, the application of electrochemical noise analysis to tribocorrosion was reviewed. Investigations involving PSD noise analyses on various tribo-electrochemical cells for passivating materials were conducted by Ponthiaux et al. [65], by Déforge et al. [51], and in more details by Berradja et al. [46, 66]. In this latter work, the noise spectra were measured on AISI 304 L stainless steel versus corundum in a Ringer's solution in a pin-on-disk tribometer under stationary sliding-corrosion regime conditions, either at open-circuit potential or at a controlled potential. The PSD of the current noise has been interpreted as resulting from the overlap of the large number of discrete repassivation transients at the contact junctions, including the double-layer charge and the strong dependence of depassivation and repassivation kinetic rates of the oxide surface film on the sliding frequency. This was consistent with the shift in the PSD plots of the current noise fluctuations by about a decade when the sliding frequency was varied from 0.1 to 1 Hz (see **Figure 5**). Similar findings were obtained via Déforge et al. [51] by dividing open-circuit potential fluctuations to the impedance of the electrode. A 1/f low-frequency noise is explained by a long-term drift of the surface conditions. Only a minor influence of the applied normal load was observed on the PSD plots, recommending reaching the limit rate of the depassivation of the oxide surface film.

Application of the noise analysis to tribocorrosion offers the feasibility to record in parallel the PSD of normal and tangential force fluctuations and their tie-in with the current noise data (see **Figure 6**). Force fluctuations show an almost flat

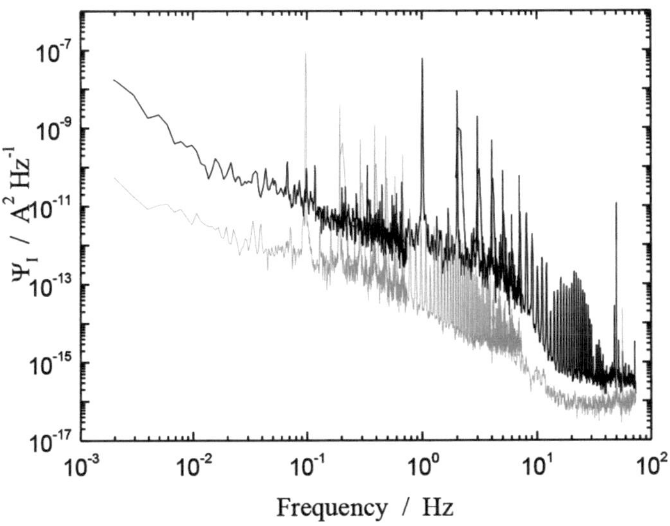

Figure 5.
PSD record of current fluctuations measured on AISI 304L during continuous sliding-corrosion test in Ringer's solution at 0.1 Hz (gray) and 1 Hz (black) and at a constant normal load of 5 N. Reproduced with permission from IOPScience [46].

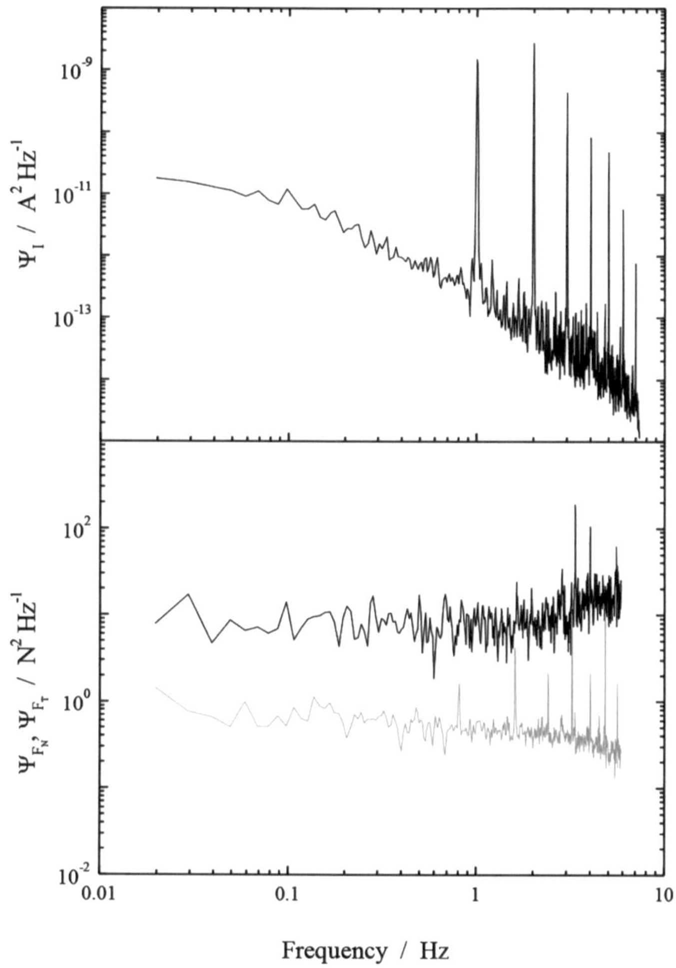

Figure 6.
PSD record of tangential (gray) and normal (black) force components measured on AISI 304L during continuous sliding-corrosion test in Ringer's solutions at a normal load of 20 N and at 1 Hz frequency. Reproduced with permission from IOPScience [46].

spectrum (white noise) as expected following short random mechanical interactions between colliding asperities, whereas the current noise is consistent with finite time-constant transient responses to the depassivation events.

1.4 Corrosion forecast by electrochemical impedance spectroscopy (EIS)

The EIS has matured greatly over the past 25 years as a tool in corrosion protection research and has proven to be one of the most useful electrochemical characterization techniques presently available. In practice, EIS has become a standardized research tool for corrosion prediction [15] and found wide applications in both fundamental and applied laboratory researches [67]. Recent applications in tribocorrosion reflect the steady progress of the EIS method in terms of research and development [62]. Compared with the LPR technique, the EIS technique is considered more advanced, since it has the ability to study high-impedance systems, in which the conventional LPR technique has failed, such as coatings and linings [16, 68], high pure water, and organic coating/metal systems [69] or corrosion in a low conductive solution [70]. This technique is especially useful for evaluating corrosion inhibitors [24, 71], analyzing the corrosion mechanisms [72, 73], and so on.

A significant number of tutorials have been addressed on the EIS experimental setup, the measurement methodology, and data analysis methods [27, 74–78]. The technique has been of a great deal of concern to the extent that an ASTM standard, i.e., ASTM G106-89, has been produced to provide the practitioner with a test method to verify that the electronic equipment, the electrochemical cell, and the spectrum generation algorithm impedance work properly [15].

1.4.1 Principle of the EIS technique

The EIS technique normally uses a typical three-electrode cell system controlled by a potentiostat, similar to that used in the LPR technique. Unlike the previous time-resolved techniques, where the current system response is either the consequence of a large voltage perturbation from the steady-state condition (case of Tafel extrapolation) or from a smaller perturbation (case of LPR method), in the EIS approach, however, by applying a small varying perturbation over a range of frequency, it is possible to probe the full response of the electrochemical system, and not just the resistive components. In that respect, a small AC signal, i.e., alternating potential or voltage $V(\omega)$ typically a sine wave of amplitude ± 10 mV of the corrosion potential, is applied over a wide range of frequency (typically from 10^5 down to 10^{-2} or 10^{-3} Hz) at a number of discrete frequencies (typically 5–10 frequencies per decade), and the alternating current response, $i(\omega)$, is measured at each frequency, ω (i.e., the ac polarization or angular frequency, $\omega = 2\pi f$). For a linear system, the current response signal will be a sine wave of the same frequency as the excitation signal (voltage) but shifted in phase. This is transmitted to a frequency response analyzer or a lock-in amplifier to calculate the impedance and phase shift. Full frequency sweeps provide phase-shift information that can be used in combination with equivalent circuit models to gain valuable information from the complex interface of the corrosion system. The frequency-dependent impedance is determined by the relation: $Z(\omega) = V(\omega)/i(\omega)$.

1.4.2 Electrode/electrolyte electrochemical interface circuit

Basically, the electrode/electrolyte interface is characterized by a separation of charges resulting in the creation of parallel planes of electrical charges whose behavior is similar to a circuit consisting of a capacitor and a resistor in parallel and certainly not to a perfect capacitor. Indeed, the current flowing in a perfect capacitor would cease when the latter would be fully charged, hence the need to add a resistor in parallel to let a weak current flow. An electrochemical interface can be viewed as an electrical circuit, or called the equivalent circuit, composed of a number of elements such as resistances (R), capacitances (C), and inductances (L) [26]. Explanations of the EIS results are usually based on the equivalent circuit used. Many software programs and packages are now available for fitting the impedance spectra to analogous circuits [15], a strategy often used to analyze data. Further information on the EIS measurements and instrumentation can be found elsewhere [15, 17, 79, 80].

Not all the available proposed equivalent circuits to model electrochemical interfaces can actually satisfy what is applied to a freely corroding system. In most cases, the impedance corresponding to a simple corrosion process, under activation control, can be represented by the well-known Randles' [81] equivalent circuit (RC circuit) which allows to describe the behavior of many electrochemical electrode/electrolyte interfaces. A typical example is shown in **Figure 7**, where R_S, R_{CT},

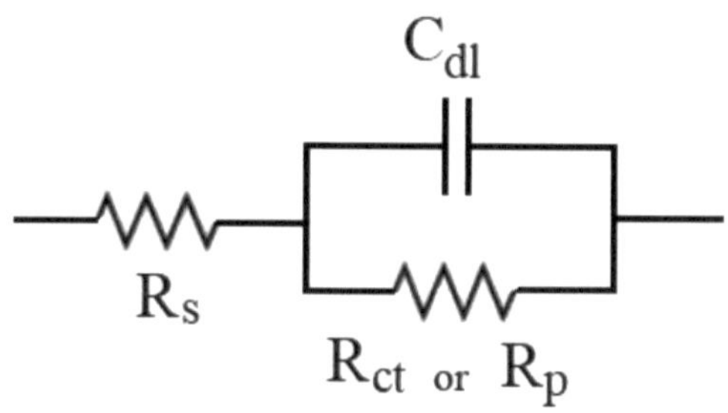

Figure 7.
A simple Randles-type equivalent circuit (RC).

and C_{DL} represent labels for the solution resistor, the Faradaic charge transfer resistor, and the double-layer capacitance, respectively. The capacity is associated with the separation of charges at the electrode/electrolyte interface as in the case of a working electrode having a surface film (e.g., AISI 304 stainless steel immersed in a 0.5 M H_2SO_4 electrolyte), in which case the capacity of the equivalent circuit can be associated with the capacity of the passive oxide surface film and the resistor in parallel with the capacitor is considered as the charge transfer resistance, R_{CT} (or the polarization resistance, R_P, under *EIS-free corrosion* conditions), while the ohmic resistance in solution, R_S, between the working electrode and the reference electrode is in series with the parallel resistor and the capacitor. If the amplitude of the perturbation signal is small enough (e.g., a voltage less than 20 mV), R_{CT} can be regarded as equivalent to the linear polarization resistance (R_P).

The behavior of such an electrochemical interface can be described by Eq. (16):

$$Z(\omega) = R_s + \frac{R_p}{1 + (j\omega R_P C_{DL})^{\beta}} \tag{16}$$

R_{CT} or R_P can be determined in several ways. A convenient way is to use the Nyquist diagram. For the simple Randles-type equivalent circuit as shown in **Figure 7**, the corresponding Nyquist diagram is displayed in **Figure 8**, in which a perfect semicircle is observed. The high-frequency response is used to determine the component of R_S involved in the measurement. R_S can be read directly from the abscissa when the angular frequency ω ($\omega = 2\pi f$) tends to be infinite (f_{max} or $f \rightarrow \infty$). The total resistance ($R_P + R_S$) can also be read from the abscissa when ω approaches zero (f_{min} or $f \rightarrow 0$). So, R_P can be determined by subtracting the R_S value from the

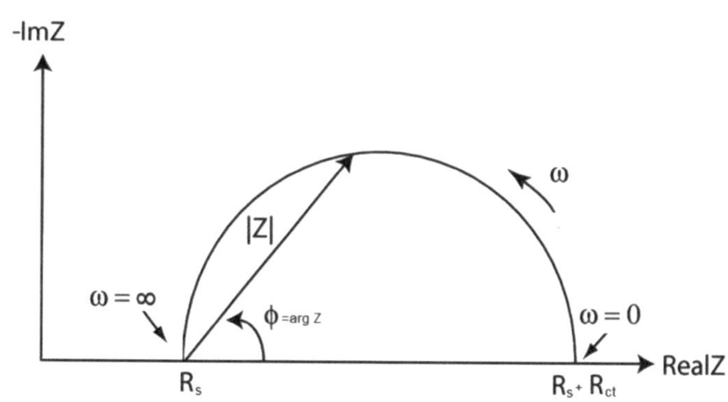

Figure 8.
The Nyquist diagram responding to the simple Randels-type equivalent circuit.

low-frequency measurement. The conversion of the polarization resistance into a corrosion rate requires an independent empirical measurement of the Tafel slopes using a potentiodynamic polarization method and/or harmonic distortion analysis or otherwise taken from the literature. The double-layer capacitance, C_{DL}, can also be determined for a system exhibiting a behavior similar to that of a perfect RC circuit from the values of R_P and the maximum frequency, f_{max}, that corresponds to the frequency of the point at which the imaginary component has a maximum value, viz.:

$$C_{DL} = \frac{1}{2\pi f_{\max} R_p} \tag{17}$$

It is worth of note that in practice, f cannot really go as high as infinite; it is inevitable that some extrapolation has to be made. Extrapolation at the high-frequency limit usually presents few issues because the impedance becomes nonreactive at frequencies as low as 10 kHz in most cases [82]. On the other hand, reactance is still commonly observed at frequencies as low as 10^{-3} Hz [82]. Therefore, special precautions must be taken to obtain reliable data and to avoid possible artifacts [17, 83]. Furthermore, the measurement cycle time depends on the frequency range used, in particular the low frequencies. For instance, a single-frequency cycle at 10^{-3} Hz needs about 15 min of testing time. A high-to-low-frequency analysis moving down to 10^{-3} Hz frequency likely requires more than 2 hours of scan time. In order to perform a normal standard corrosion monitoring with the EIS technique, assistance is needed to optimize the use of the high-frequency data and reduce measurement time. There is a constant need to improve data processing and analysis in order to minimize uncertainties and to allow the EIS technique becoming user-friendly for corrosion monitoring in both laboratory and field facilities, though it must be emphasized that the need for an easy-to-deploy field instrument has always been an obstacle to online corrosion monitoring with the EIS technique.

An alternative to the impedance model in the Nyquist diagram involves the conversion of the impedance into a complex number. The impedance can thus be designated by an amplitude, |Z|, and a phase shift, ϕ, or by the sum of the real (Z') and imaginary (Z'') components, such that,

$$Z(\omega) = Z'(\omega) + jZ''(\omega) \tag{18}$$

Both the *log*|Z| data and the phase angle ϕ are plotted against the angular frequency, *log* ω, of the excitation signal, a format which substitutes for the Nyquist diagram, i.e., the so-called Bode diagram. **Figure 9** shows how the same data (Nyquist plot) appears in a Bode plot format with respect to the equivalent circuit of **Figure 7**.

Highest (ω_H) and lowest frequencies (ω_L) can be readily determined. As shown in **Figure 9**, Z is independent of the frequency at ω_H and ω_L, limit values represented by horizontal lines. From these lines, values of R_S and (R_S + R_{CT}) can be measured. This analysis forms the basis of the corrosion monitoring as proposed by Tsuru et al. [74] to allow the determination of |Z| at each frequency in the horizontal portions of the Bode diagram.

Sometimes, it is not convenient to perform impedance measurements at very low frequencies (as in DC techniques such as linear polarization). However, it is still possible to extrapolate the polarization resistance, R_P, from the Bode diagram. In **Figure 9**, the low- and high-frequency breakpoints (i.e., ω_L and ω_H, respectively) can be determined from the 45° phase angle Bode diagram (see the pseudo-Gaussian curve). The intersection point A can be determined from the *log* ω_H

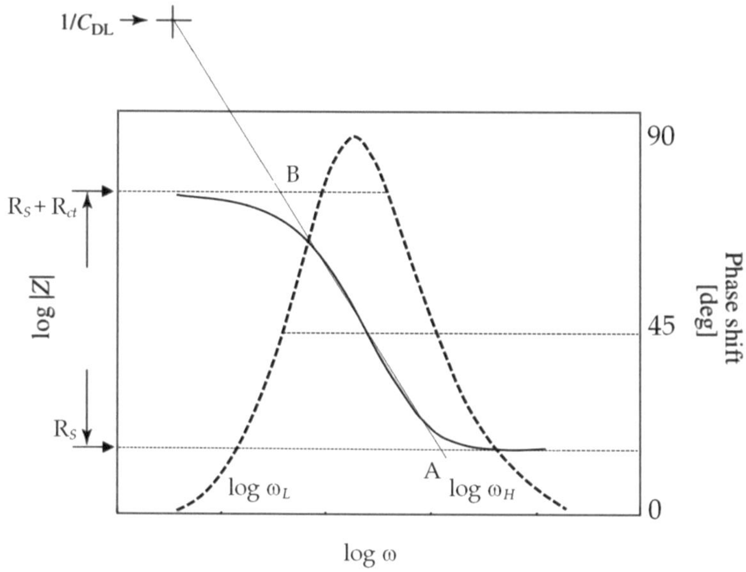

Figure 9.
Bode diagram with respect to the Randels-type equivalent circuit in ***Figure 7****.*

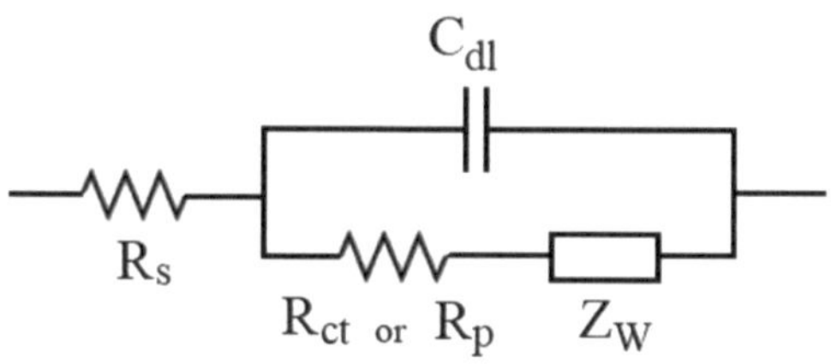

Figure 10.
A Randles-type equivalent circuit including Warburg impedance component, Z_W.

and R_S. By extrapolating from A toward the central linear portion of the |Z| curve, a linear line can be determined. On this line, point B is obtained at $log\ \omega_L$. With the projection of point B to the log|Z| axis, the total resistance ($R_S + R_{CT}$) can be measured. In this way, R_P can be determined. At intermediate frequencies, the capacitor affects the response of the overall RC circuit.

The situation struggles when diffusion processes govern the corrosion behavior. A convenient way to deal with this complication is to add a Warburg impedance. The latter describes the impedance of the concentration and diffusion processes in the equivalent circuit as shown in **Figure 10**.

The Warburg impedance, Z_W, is given by the equation

$$Z_W = \frac{\sigma_w}{\sqrt{\omega}} - j\frac{\sigma_w}{\sqrt{\omega}} \tag{19}$$

where σ_w is the Warburg coefficient.

Eq. (19) implies that, whatever the frequency, the real and imaginary parts of the Warburg impedance are equal and inversely proportional to $\sigma_w^{1/2}$. In the Nyquist plot, this impedance will result in a straight line at a constant phase angle at 45°, as shown in **Figure 10**. However, the effect of the Warburg impedance can complicate the correct estimate of the R_P value in certain cases. Therefore, the impedance data must be numerically adjusted to fit with the correct model to facilitate the extraction of the total resistance ($R_S + R_P$) from the abscissa or by using an appropriate modeling software. However, the situation can readily become more complicated if other effects, such as time-constant dispersion, adsorption processes, and so on, are taken into account; the time-constant dispersion, which can be caused by inhomogeneities in the corroded surface, results in a depression of the semicircle [75, 76]. Adsorption, on the other hand, can reveal a second semicircle at low frequencies [77]. All these effects can occur simultaneously [27], making the interpretation of impedance data rather more difficult [78, 84] (**Figure 11**).

There is a need for an appropriate model equivalent circuit beyond the existing model standards to remedy that shortcoming. An "appropriate" model is understood not only as a good fit of the impedance data but also as a rational explanation of the underlying corrosion mechanism. Moreover, the requirement of sophisticated AC frequency generator and analyzer and the time needed to acquire the complete impedance diagram (particularly in the range of low frequency) impose a serious limitation in real-time corrosion monitoring applications. Other disadvantages include a priori knowledge of the Tafel parameters in order to convert the polarization resistance into a corrosion rate and the fact that it is too difficult to detect and monitor localized corrosion, even if such applications have been explored [85, 86].

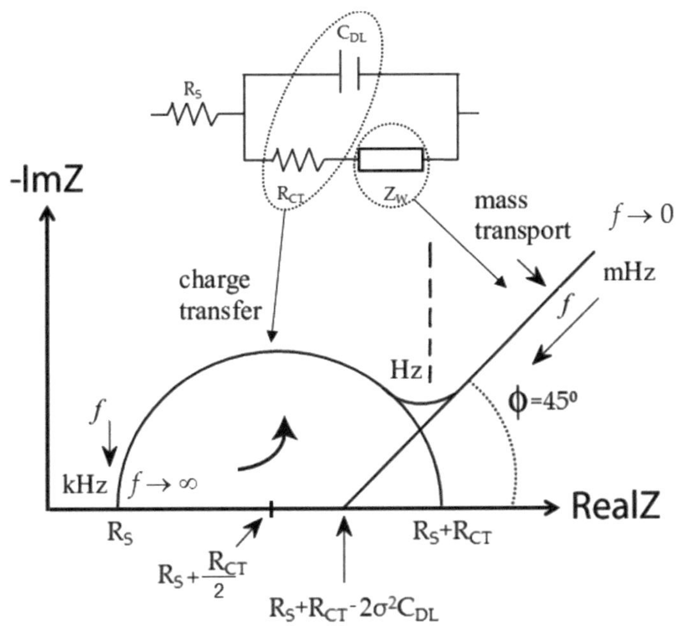

Figure 11.
The Nyquist diagram responding to the equivalent circuit of ***Figure 10****.*

1.4.3 Illustrative examples of the application of EIS in corrosion and tribocorrosion systems

Attempts were made to use the EIS technique in corrosion and corrosion-wear monitoring of Fe-31% Ni electrode immersed in 0.5 M H_2SO_4 [87]. The corresponding Nyquist impedance diagrams were recorded at an anodic potential of −675 mV/SSE (+100 mV/open-circuit potential) before and during sliding-corrosion as shown in **Figure 12**. At this potential, the prevailing reaction is dissolution. At high frequency, under free corrosion and unloaded conditions, the capacitive arc reveals the influence of the dielectric properties of the electrochemical double layer and the charge transfer due to electrochemical reactions. Under

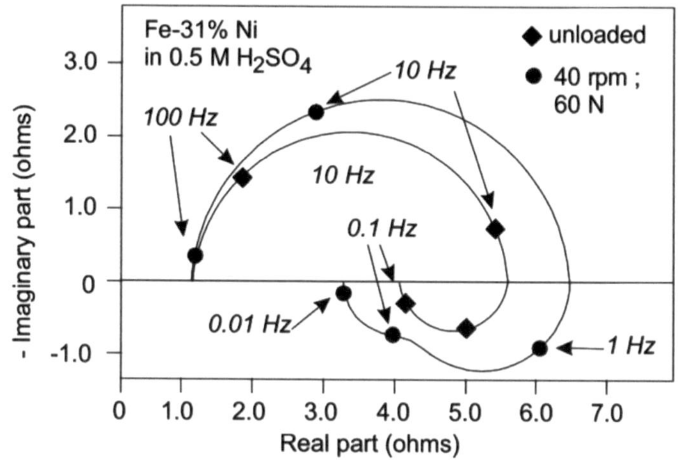

Figure 12.
Nyquist plots recorded at E = −675 mV/SSE (I = 20 mA) on Fe—31% Ni in 0.5 M H_2SO_4 under free (unloaded) and sliding conditions (against a corundum counterbody pin; 60 N normal force, sliding speed 0.031 m s^{-1}). Reproduced from [8] with permission from Elsevier.

sliding conditions, the size of the capacitive arc increases, suggesting an increase in the transfer resistance and a decrease in the reactivity of the surface, consistent with the effect of mechanical straining of the worn surface. At low frequency, however, the inductive arc indicates the relaxation of the surface concentration of adsorbed intermediate species involved in the dissolution mechanism. Under corrosion-wear conditions, the kinetics of the dissolution process is apparently modified, as revealed by the second inductive loop in the diagram. Given that not all of these investigations have been concluded, a detailed explanation is not straightforward, and further research is recommended. Although these impedance measurements provide a convenient way to study the mechanism of electrochemical reactions involved in tribocorrosion processes, still the interpretation of impedance records during sliding-corrosion experiments is rather difficult because of the heterogeneous surface-state condition. Actually, a nonuniform distribution of the electrochemical impedance on the steel surface must be taken into account. The action of friction can be analyzed thoroughly if this distribution is known. Equivalent electrical circuit models or finite element models could be used to obtain impedance distributions and to calculate the overall impedance.

2. Comparison of the techniques for the assessment of corrosion rate

The transposition of the foregoing electrochemical techniques to corrosion situations is illustrated in [63] for the assessment of corrosion rate. The results presented in **Table 1** summarize the data generated by the different techniques for Fe electrodes in 0.5 M H_2SO_4 under well-controlled conditions and their corresponding corrosion current densities, resistances, and required parameters which determined those data.

All these techniques monitor the electrode response following the stimulation by a potential variation in time or frequency domain with the exception of the electrochemical noise analysis technique. The extent of the potential stimulation and the current response decreases in the order from Tafel extrapolation method, linear polarization, EIS, to electrochemical noise. Each of these techniques provides the necessary information for a given corroding system, and there are trade-offs involved in the comparative decision of which is the best to use.

Techniques	Parameters						
	b_a [mV decade^{-1}]	b_c [mV decade^{-1}]	i_{corr} [A cm^{-2}]	C_{DL} [μF cm^{-2}]	R_P [Ω cm^2]	R_Ω [Ω cm^2]	R_n [Ω cm^2]
Linear polarization (LPR)	—	—	1.4×10^{-4}	—	80	—	—
Tafel extrapolation	34	114	1.8×10^{-4} Cathodic extrap. 7×10^{-5} Anodic extrap.	—	—	—	—
Electrochemical impedance (EIS)	—	—	1.1×10^{-4}	333 (Bode) 116 (Nyquist) 84 (fit)	98	0.3	—
Electrochemical noise (ENA)	—	—	—	—	—	—	20 –40

Reproduced from [63] with permission from Wiley Online Library.

Table 1.
Data outcomes determined by different electrochemical techniques on Fe in 0.5 M H_2SO_4.

Author details

Abdenacer Berradja
MTM Department, K.U. Leuven, Leuven, Belgium

*Address all correspondence to: a.berradja@gmail.com

References

[1] Landolt D, Mischler S. Tribocorrosion of Passive Metals and Coatings. Sawston, Cambridge: Woodhead Publishing Limited; 2011. ISBN: 978-1-84569-966-6

[2] Celis JP, Ponthiaux P, editors. Testing tribocorrosion of passivating materials supporting research and industrial innovation. In: Handbook EFC 62. UK, France, Germany, Belgium: Maney Publishing; 2012. ISBN: 9781907625202

[3] Berradja A. Metallic glasses for triboelectrochemistry systems, Chap 5. In: Huang H, editor. Metallic Glasses—Properties and Processing. London, UK: Intech Open; 2018. ISBN: 978-953-51-6208-7

[4] Landolt D. Corrosion and Surface Chemistry of Metals. EFPL Press; 2007. pp. 415-460. Chap 10

[5] Frankel GS. Electrochemical techniques in corrosion: Status, limitations, and needs. Journal of ASTM International. 2008;5:101241. DOI: 10.1520/JAI101241

[6] Trenthewey KR, Camberlain J. Corrosion for Science and Engineering. 2nd ed. United States: Longman Group Limited; 1995. ISBN-13: 978-0582238695

[7] Roberge PR. Handbook of Corrosion Engineering. New York, Chicago, San Francisco, Lisbon, London, Madrid, Mexico City, Milan, New Delhi, San Juan, Seoul, Singapore, Sydney, Toronto: McGraw-Hill Education; 2000. ISBN: 007-076516-2

[8] Haruyama S, Tsuru T. A corrosion monitor based on impedance method. In: Mansfeld F, Bertocci U, editors. Electrochemical Corrosion Testing, ASTM STP 727. Barr Harbor Drive, West Conshohocken, PA: American Society for Testing and Materials; 1981. pp. 167-186

[9] Epelboin I, Gabrielli C, Keddam M, Takenouti H. Alternating-current impedance measurements applied to corrosion studies and corrosion-rate determination. In: Mansfeld F, Bertocci U, editors. ASTM STP 727. Barr Harbor Drive, West Conshohocken, PA: American Society for Testing and Materials; 1981. p. 110

[10] ASTM G59-97(2014). Standard Test Method for Conducting Potentiodynamic Polarization Resistance Measurements. West Conshohocken, PA: ASTM International; 2014

[11] Mansfeld F, Xiao H. Electrochemical Noise Measurements for Corrosion Applications, ASTM STP 1277. Philadelphia, PA: American Society for Testing and Materials; 1996

[12] Dawson JL. Electrochemical noise measurement: The definitive in-situ technique for corrosion applications? In: Kearns JR, Scully JR, Roberge PR, Reichert DL, Dawson JL, editors. Electrochemical Noise Measurement for Corrosion Applications. ASTM STP 1277. pp. 3-35

[13] Kearns JR, Eden DA, Yaffe MR, Fahey JV, Reichert DL, Silverman DC. ASTM standardization of electrochemical noise measurement, ASTM STP 1277. In: Kearns JR, Scully JR, Roberge PR, Reichert DL, Dawson JL, editors. Electrochem. Noise Meas. Corros. Appl. Philadelphia, PA: ASTM; 1996. pp. 446-470

[14] Reichert DL. Electrochemical noise measurement for determining corrosion rates. In: Kearns JR, Scully JR, Roberge PR, Reichert DL, Dawson JL, editors. Electrochem. Noise Meas. Corros. Appl. Philadelphia, PA: ASTM; 1996. pp. 79-92

[15] ASTM Standard: G106-89, Standard Practice for Verification of Algorithm and Equipment for Electrochemical Impedance Measurements, Annual Book of ASTM Standards. Wear and Erosion, Metal Corrosion. Vol. 03.02. West Conshohocken, PA: ASTM; 2001

[16] Kamarchik. In: Scully JR, Silverman DC, Kendig MW, editors. Electrochemical Impedance: Analysis and Interpretation, ASTM STP 1188. Philadelphia: American Society for Testing and Materials; 1993. p. 463474

[17] Macdonald DD, McKubre MCH. Electrochemical impedance techniques in corrosion science. ASTM STP 727. 1981:110-149

[18] Bonhoeffer KF, Jena W. Über das elektromotorische Verhalten von Eisen. Zeitschrift für Elektrochemie und angewandte physikalische Chemie. 1951;**55**:151-154

[19] Wagner C, Traud W. Über die Deutung von Korrosionsvorgängen durch Überlagerung von elektrochemischen Teilvorgängen und über die Potentialbildung an Mischelektroden. Zeitschrift für Elektrochemie und angewandte physikalische Chemie. 1938;**44**:391-402

[20] Mansfeld F. The polarization resistance technique for measuring corrosion currents (Chap 3). In: Fontana MG, Staehle RW, editors. Advances in Corrosion Engineering and Technology. Vol. 6. New York: Plenum; 1976

[21] Stern M, Geary AL. Electrochemical polarization. Journal of the Electrochemical Society. 1957;**104**:56

[22] Application Note Corr-1, Basics of Corrosion Measurements, EG&G Princeton Applied Research, 1982

[23] Mansfield F. Tafel slopes and corrosion rates from polarization resistance measurements. Corrosion. 1973;**29**:397

[24] Lorenze WJ, Mansfeld F. Determination of corrosion rates by electrochemical DC and AC methods. Corrosion Science. 981;**21**:647-672

[25] Mansfeld F, Oldham KB. A modification of the Stern—Geary linear polarization equation. Corrosion Science. 1971;**11**:787

[26] Hladky K, Callow LM, Dawson JL. Corrosion Rates from Impedance Measurements: An Introduction. British Corrosion Journal. 1980;**15**:20

[27] Epelboin I, Keddam M, Takenouti H. Use of impedance measurements for the determination of the instant rate of metal corrosion. Journal of Applied Electrochemistry. 1972;**2**:71-79

[28] Boukamp BA. A non-linear least squares fit procedure for analysis of immittance data of electrochemical systems. Solid State Ionics. 1986;**20**: 31-44

[29] Chen JF, Bogaerts WF. Electrochemical emission spectroscopy for monitoring uniform and localized corrosion. Corrosion. 1996;**52**:753-759

[30] Wu P-Q, Quan ZL, Celis J-P. On-line corrosion and corrosion-wear monitoring using a modified electrochemical noise technique. Materials and Corrosion-Werkstoffe und Korrosion. 2005;**56**(6):379

[31] Bertocci U, Gabrielli C, Huet F, Keddam M, Rousseau P. Noise resistance applied to corrosion measurements II. Experimental tests. Journal of the Electrochemical Society. 1997;**144**:37-43

[32] Bertocci U, Gabrielli C, Huet F, Keddam M. Noise resistance applied to corrosion measurements. I. Theoretical

analysis. Journal of the Electrochemical Society. 1997;**144**:31-37

[33] Eden DA, Hladky K, John DG, Dawson JL. CORROSION'86, paper 274. 1986

[34] Tafel J. Uber die Polarisation bei kathodischer Wasserstoffentwicklung Zeitschrift für physikalische Chemie. Stochiometrie und Verwandtschaftslehre. 1905;**50**:641

[35] Papavinasam S. Electrochemical polarization techniques for corrosion monitoring (chap). In: Yang L editor. Techniques for Corrosion Monitoring. Sawston, Cambridge: Woodhead Publishing; 2008. ISBN: 9781845691875

[36] ASTM G5-14. Standard Reference Test Method for Making Potentiodynamic Anodic Polarization Measurements. West Conshohocken, PA: ASTM International; 2014

[37] Hladky K, Dawson JL. The measurement of localized corrosion using electrochemical noise. Corrosion Science. 1981;**21**:317

[38] Hladky K, Dawson JL. The measurement of corrosion using electrochemical 1/f noise. Corrosion Science. 1982;**22**:231

[39] Searson PC, Dawson JL. Analysis of electrochemical noise generated by corroding electrodes under open-circuit conditions. Journal of the Electrochemical Society. 1988;**135**:1908

[40] Mansfeld F, Xiao H. Electrochemical noise analysis of iron exposed to nacl solutions of different corrosivity. Journal of the Electrochemical Society. 1993;**140**:2205

[41] Bertocci U, Huet F. Noise resistance applied to corrosion measurements. III. Influence of the instrumental noise on the measurements. Journal of the Electrochemical Society. 1997;**144**: 2786-2793

[42] Bertocci U, Frydman J, Gabrielli C, Huet F, Keddam K. Analysis of electrochemical noise by power spectral density applied to corrosion studies: Maximum entropy method or fast fourier transform?. Journal of the Electrochemical Society. 1998;**145**:2780

[43] Cottis RA. Interpretation of electrochemical noise data. Corrosion. 2001;**57**(3):265-285. DOI: 10.5006/1.3290350

[44] Coleman R, editor. Stochastic Processes. Springer, Dordrecht Publisher; 1974. ISBN: 978-0-04-519017-1. DOI: 10.1007/978-94-010-9796-3

[45] Iverson WP. Transient voltage changes produced in corroding metals and alloys. Journal of the Electrochemical Society. 1968;**115**:617

[46] Berradja A, Déforge D, Nogueira RP, Ponthiaux P, Wenger F, Celis JP. An electrochemical noise study of tribocorrosion processes of AISI 304L in Cl^- and SO_4^{2-} media. Journal of Physics D: Applied Physics. 2006;**39**:3184-3192

[47] Wu P-Q, Celis J-P. Electrochemical noise measurements on stainless steel during corrosion-wear in sliding contacts. Wear. 2004;**256**:480-490

[48] Dawson JL, Hladky K, Eden DA. Electrochemical noise—Some new developments in corrosion monitoring, UK Corrosion, 83. In: Proceedings of the Conference. Exeter House, 48 Holloway Head, Birmingham, UK: The Institution of Corrosion Science and Technology; 1983. pp. 99-108

[49] Bertocci U, Mullen JL, Ye Y-X. Electrochemical noise measurement for the study of localized corrosion and passivity breakdown. In: Froment M, editor. Passivity of Metals and

Semiconductors. Amsterdam: Elsevier Science; 1983

[50] Hladky K, Lomas JP, John DG, Eden DA, Dawson JL. Corrosion monitoring using electrochemical noise: Theory and practice. In: Corrosion Monitoring and Inspection in the Oil, Petrochemical and Process Industries, Oyez Scientific and Technical Services. London: Bath House. p. 1984

[51] Déforge D, Huet F, Nogueira RP, Ponthiaux P, Wenger F. Electrochemical noise analysis of tribocorrosion processes under steady-state friction regime. Corrosion. 2006;**62**:514-521

[52] Leban M, Dolecek V, Legat A. Comparative analysis of electrochemical noise generated during stress corrosion cracking of AISI 304 stainless steel. Corrosion. 2000;**56**:921-927

[53] Watanabe Y, Kondo T. Current and potential fluctuation characteristics in intergranular stress corrosion cracking processes of stainless steels. Corrosion. 2000;**56**:1250-1255

[54] Benish ML, Sikora J, Shaw B, Sikora E, Yaffe M, Krebs A, et al. A new electrochemical noise technique for monitoring the localized corrosion of 304 stainless steel in chloride-containing solutions. Corrosion '98. Paper No. 370

[55] Berradja A, Bratu F, Benea L, Willems G, Celis J-P. Effect of sliding wear on tribocorrosion behaviour of stainless steels in a Ringer's solution. Wear. 2006;**261**:987-993

[56] Mohrbacher H, Celis JP, Roos JR. Laboratory testing of displacement and load induced fretting. Tribology International. 1995;**28**:269-278

[57] Bertocci U, Kruger J. Studies of passive film breakdown by detection and analysis of electrochemical noise. Surface Science. 1980;**101**:608

[58] Eden A, John DG, Dawson JL. International Patent W087/070222

[59] Ritter S, Huet F, Cottis RA. Guideline for an assessment of electrochemical noise measurement devices. Materials Science. 2012;**63**: 297-302. DOI: 10.1002/maco.201005839

[60] Lengyel B, Mészáros L, Mészáros G, Fekete E, Janaszik F, Szenes I. Electrochemical methods to determine the corrosion rate of a metal protected by a paint film. Progress in Organic Coating. 1999;**36**:11-14

[61] Xiao H, Mansfeld F. Evaluation of coating degradation with electrochemical impedance spectroscopy and electrochemical noise analysis. Journal of the Electrochemical Society. 1994;**141**:2332

[62] Keddam M, Ponthiaux P, Vivier V. Tribo-electrochemical impedance: A new technique for mechanistic study in tribocorrosion. Electrochimica Acta. 2013

[63] Frankel GS, Rohwerder M. Electrochemical techniques for corrosion. In: Encyclopedia of Electrochemistry. Wiley VCH: Weinheim Germany; 2007

[64] Oltra R, Gabrielli C, Huet F, Keddam M. Electrochemical investigation of locally depassivated iron. A comparison of various techniques. Electrochimica Acta. 1986; **31**(12):1505-1511. DOI: 10.1016/0013-4686(86)87068-2

[65] Ponthiaux P, Wenger F, Galland J, Lederer G, Celati N. Du bruit électrochimique pour déterminer la surface dépassivée par frottement cas d'un acier z2cnd 17–13 en milieu chloruré (NaCl 3%), Matériaux & Techniques – numéro hors série1997. pp. 43-46

[66] Celis JP, Berradja A, Deforge D, Wenger F, Ponthiaux P, Nogueira RP.

Tribocorrosion of AISI 304L stainless steel under continuous sliding in Ringer's medium. In: Conference on Chemical, Electrochemical, and Mechanical Effects on CMP, Tribocorrosion, and Biotribocorrosion, 207th Meeting; Quebec City, Canada; May 15-20. 2005

[67] Gabrielli C, Keddam M. Review of applications of impedance and noise analysis to uniform and localized corrosion. Corrosion. 1992;**48**:794-811

[68] Beavers A, Thompson NG, Silverman DC. Corrosion 93, Paper No. 348

[69] EG&G. Evaluation of Organic Coatings by Electrochemical Impedance Measurements, Application Note AC-2. 1990

[70] Mansfeld F, Kendig MW. Determination of the polarization resistance from impedance measurements. Materials and Corrosion Werkstoffe und Korrosion. 1983;**34**:397

[71] Mansfeld F, Kendig MW, Lorenz WJ. Corrosion inhibition in neutral, aerated media. Journal of the Electrochemical Society. 1985;**132**:290

[72] Kilpatrick JM. Measuring corrosion rate-now. Oil and Gas Journal. 1964;**62**:155

[73] Epelboin I, Wiart R. Mechanism of the electrocrystallization of nickel and cobalt in acidic solution. Journal of the Electrochemical Society. 1971;**118**:1577

[74] Tsuru T, Haruyama S, Gijutsu B. Corrosion inhibition of iron by amphoteric surfactants in 2M HCl. Japan Society of Corrosion Engineering. 1978;**27**:573

[75] Cole SK, Cole RH. Dispersion and absorption in dielectrics I. Alternating current characteristics. The Journal of Chemical Physics. 1941;**9**:341

[76] Epelboin I, Keddam M. Faradaic impedances: Diffusion impedance and reaction impedance. Journal of the Electrochemical Society. 1970;**117**:1052

[77] Armstsrong RD, Henderson M. Impedance plane display of a reaction with an adsorbed intermediate. Journal of Electroanalytical Chemistry. 1972;**39**:81

[78] Mansfeld F. Recording and analysis of ac impedance data for corrosion studies. Corrosion. 1981;**37**:301

[79] EG&G. Basics of AC Impedance Measurements, Application Note AC-1. 1990

[80] Hubrecht J. In: Helsen JA, Breme HJ, editors. Chap 14. Metals as Biomaterials. Electrochemical Impedance Spectroscopy as a Surface Analytical Technique for Biomaterials. John Wiley & Sons Ltd.; 1998. pp. 405-466

[81] Randles EB. Kinetics of rapid electrode reactions. Discussions of the Faraday Society. 1947;**1**:11

[82] Macdonald DD. Some advantages and pitfalls of electrochemical impedance spectroscopy. Corrosion. 1990;**46**(3):229-242

[83] Syrett C, Macdonald DD. The validity of electrochemical methods for measuring corrosion rates of copper-nickel alloys in sea water. Corrosion. 1979;**35**:505

[84] Titz J, Wagner GH, Spähn H, Ebert M, Jüttner K, Lorenz WJ. Characterization of organic coatings on metal substrates by electrochemical impedance spectroscopy. Corrosion. 1990;**46**:221

[85] Mansfeld F, Lin S, Kim S, Shih H. Pitting and surface modification of SIC/Al. Corrosion Science. 1987;**27**:997

[86] Vera P, Nishikata CA, Tsuru T. AC impedance monitoring of pitting corrosion of stainless steel under a wet-dry cyclic condition in chloride-containing environment. Corrosion Science. 1996;**38**:1397

[87] Ponthiaux P, Wenger F, Drees D, Celis JP. Electrochemical techniques for studying tribocorrosion processes. Wear. 2004;**256**:459-468